W0016679

SpringerBriefs in Agriculture

For further volumes:
http://www.springer.com/series/10183

Muhammad Asif · Muhammad Iqbal
Harpinder Randhawa · Dean Spaner

Managing and Breeding Wheat for Organic Systems

Enhancing Competitiveness Against Weeds

 Springer

Muhammad Asif
Dean Spaner
Agricultural, Food and Nutritional Science
University of Alberta
Edmonton, AB
Canada

Harpinder Randhawa
Agriculture and Agri-Food Canada
Lethbridge Research Centre
Lethbridge, AB
Canada

Muhammad Iqbal
National Agricultural Research Centre
Islamabad
Pakistan

ISSN 2211-808X   ISSN 2211-8098 (electronic)
ISBN 978-3-319-05001-0   ISBN 978-3-319-05002-7 (eBook)
DOI 10.1007/978-3-319-05002-7
Springer Cham Heidelberg New York Dordrecht London

Library of Congress Control Number: 2014933790

Printed on acid-free paper

Springer is part of Springer Science+Business Media (www.springer.com)

Preface

Breeding for organic agriculture is gaining enormous attention in the scientific community due to burgeoning trade, production, and consumption of organic produce. It is well established that organically managed lands represent a different environment than conventionally managed lands mainly due to the presence of large weeds populations. Wheat is one of the most produced and consumed cereal grains worldwide. Weed infestation is a ubiquitous threat to the production of major cereal crops including wheat (*Triticum aestivum* L.), maize (*Zea mays* L.), and rice (*Oryza sativa* L.). This threat needs to be minimized in order to maximize global crop production and to meet the food demand of the ever-increasing human population. Currently, farmers control weed infestation mainly through the application of herbicides. According to Transparency Market Research, herbicide is the largest category in the crop protection sector that contributed about $19.9 billion in 2011. The use of herbicides is increasing on an annual basis. The continuous exposure of weed species/plants to strong herbicides has inevitably resulted in the development of herbicide-resistant weed populations in at least 80 crops in 63 countries (http://www.weedscience.com), which poses a major threat/challenge to global food production and security. This widespread evolution of herbicide-resistant weed species necessitates the development of alternate weed control strategies. Therefore, interest in breeding cereals, especially wheat, to enhance competitive ability is growing among the scientific community. Various traits of economic importance, conferring competitive ability against weeds, have been identified along with improved understanding of allelopathy. The combined effects of competition and allelopathy offer a great potential to achieve maximum weed suppression. Breeding efforts have resulted in the development of weed suppressive rice cultivars that are commercially available in China and the USA, whereas research work is being conducted in many parts of the world, including Canada, to develop a highly competitive wheat ideotype.

In planning this monograph, our main intent was to (i) describe and critically review key aspects of breeding wheat for organically managed systems to enhance competitive ability against weeds and (ii) to provide a unique and time-based resource for plant breeders, agronomists, teachers, students, and weed scientists around the globe to seek information on a discipline of crop competitiveness. Consequently, this brief is divided into five chapters which are arranged in logical progression. Chapter 1 highlights the importance, history, production, and utilization of wheat from a global perspective. Chapter 2 begins with a brief overview of organic agriculture and extends a comprehensive discussion on crop–weed competitiveness. It also identifies traits of interests in different crops to breed for competitiveness and presents trait association to competitive ability in crop plants. Various strategies to control weed infestation and to enhance competitive ability through management, genetics, or genomic approaches have been presented in Chap. 3. It highlights various molecular studies undertaken to identify molecular markers linked with various traits conferring competitive ability in cereal crops. Chapter 4 provides strategies to breed wheat for disease resistance, quality, allelopathy, and earliness for organic systems. Chapter 5 summarizes the brief and outlines studies conducted during the last 5 years to examine competitive ability in various cereal crops throughout the world.

The brief is presented in a logical format making it available to a wide range of readers including plant breeders, agronomists, weed scientists, graduate, and undergraduate students involved in the field of agriculture and related disciplines, to help them in devising breeding strategies to deal with the problem of weed/weed infestation by enhancing/improving crop competitiveness.

Contents

1 Wheat: The Miracle Cereal 1
   1.1 Importance 1
   1.2 History. 3
   1.3 Production 3
   1.4 Utilization 4
   References 5

2 Crop Competitiveness 9
   2.1 Importance of Crop Competitiveness 10
   2.2 What is Crop Plant Competition? 12
   2.3 Association of Plant Traits to Competitiveness 13
   References 17

3 Strategies to Enhance Competitive Ability 21
   3.1 Management. 21
      3.1.1 Biological Control. 22
      3.1.2 Cultural Control 22
      3.1.3 Mechanical Weed Control 30
   3.2 Genetics and Genomics of Competitiveness 31
   References 45

4 Breeding Wheat for Organic Agriculture 53
   4.1 Breeding for Disease Resistance 53
   4.2 Breeding for Quality 54
   4.3 Breeding for Allelopathy 55
   4.4 Breeding for Early Maturity 57
   References 61

5 Conclusion. 65
   References 66

Author Biographies. 71

Index 73

Abbreviations

AMs	Arbuscular Mycorrhizas
BX	Benzoxazinoids
CPSW	Canada Prairie Spring Wheat
CSA	Canada Seed Act
CWRS	Canada Western Red Spring
DH	Doubled Haploid
DNA	Deoxyribonucleic acid
FAO	Food and Agriculture Organization
FHB	Fusarium Head Blight
FPPA	Federal Plant Protection Act
GFD	Grain Fill Duration
GFR	Grain Fill Rate
GMO	Genetically Modified Organisms
IWM	Integrated Weed Management
IWMS	Integrated Weed Management Strategy
LAI	Leaf Area Index
MAB	Marker Assisted Breeding
MAS	Marker Assisted Selection
NUE	Nutrient Use Efficiency
NWCA	Noxious Weed Control Act
PAR	Photosynthatically Active Radiation
PCR	Polymerase Chain Reaction
PGPR	Plant Growth Promoting Rhizosphere
QTL	Quantitative Trait Loci
RAPD	Random Amplified Polymorphic DNA
RFLP	Restriction Fragment Length Polymorphism
RIL	Recombinant Inbred Line
RUE	Radiation Use Efficiency
SNP	Single Nucleotide Polymorphism
SSR	Simple Sequence Repeat
UN	United Nations
US	United States
WHO	World Health Organization

Chapter 1
Wheat: The Miracle Cereal

Abstract Wheat is one of the most important cereal crops. It covers the largest area under any single crop in the world. It feeds about 40 % of the world's population and provides 20 % of the caloric and protein requirements in human nutrition. Wheat also occupies a central position in maintaining world's food security. Following incorporation of semi-dwarfing genes, wheat production doubled in the 1960s, an era called the Green Revolution. The Green Revolution resulted in the development of semi dwarf wheat cultivars that were highly responsive to inorganic fertilizer application, were early maturing and resistant to lodging. Semi dwarf cultivars also remained resistant to various diseases for many decades. Wheat genetic gains are less than 1 % per annum which are not sufficient to meet the future food demand of ever increasing human population. This chapter addresses importance, history, production and utilization of wheat from different perspectives.

Keywords Allohexaploid · Grain texture · Hardness · Protein · Starch · *Triticum turgidum* · *Triticum aestivum*

1.1 Importance

Bread or common wheat (*Triticum aestivum* L.) is one of the most important cereal crops in the world and ranks third in production after maize and rice. Wheat covers 17 % of the global crop acreage (217 million ha) with production of 675 million tons (Fig. 1.1). China is the largest producer of wheat with 117.4 million tons, followed by India producing 86.9 million tons (Fig. 1.2). The yield and production of wheat crop has been substantially increased during the last 50 years, whereas the area under wheat almost remained constant (Fig. 1.1). Bread wheat feeds about 40 % of the world's population and provides 20 % of the caloric and protein requirements in human nutrition (Gupta et al. 2005). For technical purposes wheat has been divided into different classes that include hard, medium, and soft (based on grain hardness); white, amber, or red (based on bran color) and;

M. Asif et al., *Managing and Breeding Wheat for Organic Systems*,
SpringerBriefs in Agriculture, DOI: 10.1007/978-3-319-05002-7_1,
© The Author(s) 2014

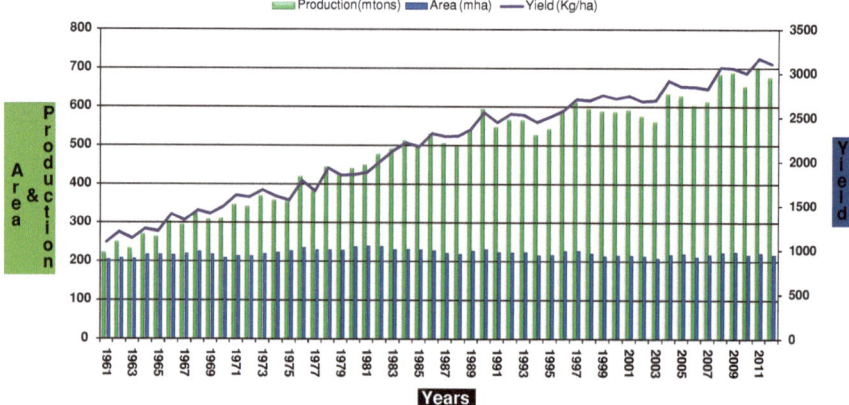

Fig. 1.1 Area, production, and yield of wheat from 1961 to 2012 (*Source* faostat.fao.org)

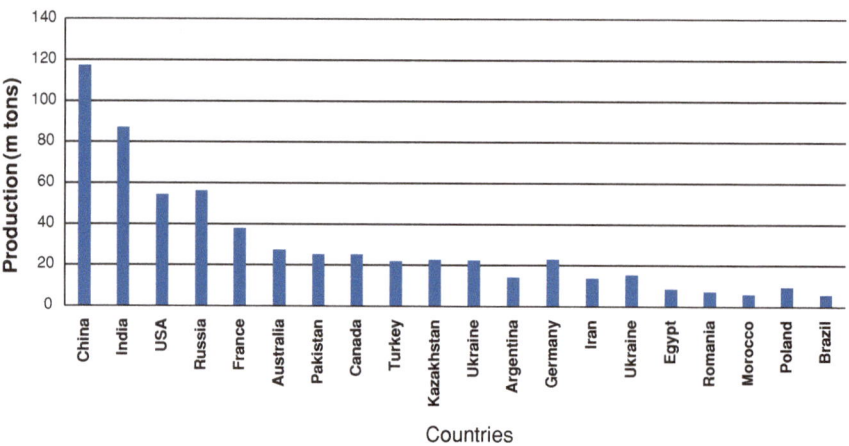

Fig. 1.2 Top 20 wheat producing countries in the world (*Source* faostat.fao.org)

spring or winter (based on growth habit). Wheat provides 10–20 % of the daily caloric requirements to people in over 60 countries of the world. A matured grain of wheat consists of about 82.5 % endosperm, 15 % bran, and 2.5 % germ. Bread wheat provides more than 50 % of the total calories and 60 % of the total protein eaten by mankind (Sial et al. 2005). Each 100 g of wheat seed consists of about 12.6–15.4 g protein, 68–71 g carbohydrate, 66.8 g starch, 1.5–1.9 g total fat, 3.2–3.6 mg iron, and 12 g of nutritional fiber (Kumar et al. 2011). There are many types of wheat. The most common species are *Triticum aestivum* L. (also known as bread wheat) and *Triticum turgidum* or durum wheat. Bread wheat accounts for 95 % of the total wheat consumed worldwide while the latter is used mainly for making pasta and semolina products. Wheat has two major growth habits; winter wheat is sown in the fall, requires a vernalization treatment, and is harvested in the spring or summer (9 months cycle); whereas spring wheat is planted in spring and harvested in late summer or early fall (3–4 months cycle).

1.2 History

Wheat is a monocot plant species and belongs to the grass family *Poaceae*. It is a self-pollinated and a natural allohexaploid ($2n = 6x = 42$; AABBDD), containing three distinct but genetically related (homoeologous) genomes, A, B, and D, each with a haploid set of seven chromosomes. Wheat originated from hybridization between the tetraploid wild emmer wheat *Triticum turgidum* ssp. *dicoccum* ($2n = 4x = 28$; AABB) and the diploid wild wheat *Aegilops tauschii* ($2n = 2x = 14$; DD). The haploid nuclear genome of *T. aestivum* contains 16.72×10^9 bp of DNA, distributed over 42 chromosome arms (Gupta et al. 2005). The A genome of wheat ($2n = 14$) has been contributed by *Triticum monococcum*; the D genome ($2n = 14$) by *Aegilops squarrosa* (Morris and Sears 1967), and most likely the B genome by *Ae. speltoides* ($2n = 14$, BB). However, it is likely that B genome progenitor no longer exists in its original form (Yan et al. 1998). Hexaploid wheat, like many other allopolyploid plant species, exhibits a diploid-like meiotic behavior to circumvent the formation of multivalent associations of more than two homologous or homoeologous chromosomes at meiosis, which can result in genetically unbalanced gametes (Naranjo and Corredor 2004). This diploid-like chromosome pairing is genetically controlled by the *Ph1* (Pairing homoeologous) gene, located on chromosome 5BL (Riley and Chapman 1958). The *Ph1* gene, by suppressing pairing of homoeologous chromosomes, validates the assumption of disomic inheritance in genetic studies of hexaploid wheat.

1.3 Production

Following incorporation of semi-dwarfing genes, wheat production doubled in the 1960s, an era called the Green Revolution. The Green Revolution resulted in the development of semi-dwarf varieties that were highly responsive to inorganic fertilizer application, were early maturing, and resistant to lodging. Semi-dwarf varieties also remained resistant to stem rust (a serious disease of wheat) for many decades. The present wheat production in South Asia is around 95 million tons and the estimated demand for 2020 is around 137 million tons (Chatrath et al. 2007). Currently, it seems unlikely to obtain another huge sudden yield jump as occurred in the 1960s. Thus, strategies need to be developed to increase wheat yield gradually (Rajaram 2001; Singh et al. 2007). Some yield limiting factors include delayed planting, drought and heat stress, poor soil nutrition, salinity, lack of competitiveness, wheat stem rust (*Puccinia graminis* f. sp. *tritici*), leaf rust

(*Puccinia triticina*), stripe rust (*Puccinia striiformis* f. sp. *tritici*), fusarium head blight (*Fusarium graminearum*), loose smut (*Ustilago tritici*), powdery mildew (*Blumeria graminis*), common bunt (*Tilletia tritici*), and root rot (*Cochliobolus sativus*).

1.4 Utilization

The three main parts of wheat grain are separated during milling. The hard outer cover of the seed is called bran, which is also known as the grain's "skin". Germ contains the embryo of the seed that sprouts and grows into a new plant. Endosperm is the largest (about 83 %) component of the kernel. Endosperm proteins are essential for bread making and other end-use products (Nakamura 2000; Rogers et al. 2001; Shewry et al. 1992). It is the food source for the sprouting new plant. The primary determinant of wheat processing and industrial quality is grain texture or hardness that affects many end-use properties of wheat (Feiz et al. 2009). The hardness of wheat grain is chiefly controlled by Hardness locus (*Ha*), which is positioned on chromosome 5DS (Campbell et al. 1999; Law et al. 1978; Mattern et al. 1973; Ram et al. 2002). *Puroindoline b* and *Puroindoline a* (*Pinb* and *Pina*, respectively) genes lie on the *Ha* locus and are involved in grain texture (Giroux and Morris 1998; Giroux et al. 2000; Morris 2002; Wanjugi et al. 2007). The wild types of both genes, *Pinb-D1a* and *Pina-D1a*, give soft texture (*Ha*), whereas mutations in any of the two genes confer hardness (Giroux and Morris 1997, 1998). Soft wheat has reduced particle-size distribution, including many free starch granules (Devaux et al. 1998). Soft textured and white grained wheat varieties have relatively pale and starchy kernels and are preferred in manufacturing biscuits, piecrust, and breakfast foods (Knott et al. 2009). The high quantity of gluten in flour from hard wheat is preferred for making bread and fine cakes (Morris 2002). Durum (*T. durum*) is the hardest wheat and is used for manufacturing macaroni, spaghetti, and other pasta products (Morris 2002).

Grain protein content determines the nutritional value of wheat grain, as well as the rheological and technological properties of wheat flour (Zhao et al. 2010). The wheat quality of bread making is associated with the absence or presence of specific proteins and their subunits (Dhaliwal et al. 1994; Payne et al. 1987; Snape et al. 1993). In addition, bread making quality also depends on the proportion of polymeric and monomeric proteins, and the amount and size distribution of the former (Gupta et al. 1993). The endosperm proteins have a key role in the determination of wheat quality. The four major types of endosperm proteins in wheat include Prolamines, Albumins, Gliadins, and Glutenins (Gupta et al. 1992). Payne et al. (1987) reported that glutenin protein content and composition control most of the variation in wheat flour quality. Glutenins account for 80 % of wheat proteins and are the principal components that determine dough quality (Payne et al. 1987). Gliadins and glutenins are wheat storage proteins and are the principal components of wheat gluten. Gluten proteins give unique viscoelastic properties to the wheat flour. Glutenins are polymeric proteins with disulphide bonds connecting the individual glutenin subunits. Glutenin subunits are further subdivided into low molecular weight (LMW-GS) and high molecular weight (HMW-GS) subunits. LMW-GS have molecular weight of 23–68 k Da, whereas HMW-GS are from 77 to 160 k Da. In addition to molecular weight, these two subunits also differ from each other in their structure and amino acid composition (Branlard et al. 1989).

Starch is an important constituent of wheat flour (Nakamura et al. 2002; Smith et al. 1995). Starch is found in the amyloplast in the form of granules and accounts for 65–70 % of the grain (Li et al. 1999). Wheat starch consists of 70–80 % amylopectin and 20–30 % amylose (Li et al. 1999). Waxy (*Wx*) proteins are the products of waxy genes and are also known as granule-bound starch synthase. These proteins take part in the formation of amylose in kernel starch granules and amylopectin in non-storage tissue (Denyer et al. 1996; Nakamura et al. 2002; Vrinten and Nakamura 2000). Waxy genes, *Wx-B1, Wx-D1, and Wx-A1*, are found on the three homoeologous genomes of wheat. Among these genes, *Wx-A1* is present on the short arm of 7A chromosome, *Wx-B1* on the long arm of chromosome 4A, and *Wx -D1* on the short arm of chromosome 7D (Chao et al. 1989; Miura et al. 1994).

References

Branlard G, Autran J, Monneveux P (1989) High molecular weight glutenin subunit in durum wheat (*T. durum*). Theor Appl Genet 78:353–358

Campbell KG, Bergman CJ, Gualberto DG, Anderson JA, Giroux MJ, Hareland G, Fulcher RG, Sorrells ME, Finney PL (1999) Quantitative trait loci associated with kernel traits in a soft× hard wheat cross. Crop Sci 39:1184–1195

Chao S, Sharp P, Worland A, Warham E, Koebner R, Gale M (1989) RFLP-based genetic maps of wheat homoeologous group 7 chromosomes. Theor Appl Genet 78:495–504

Chatrath R, Mishra B, Ferrara GO, Singh S, Joshi A (2007) Challenges to wheat production in South Asia. Euphytica 157:447–456

Denyer K, Clarke B, Hylton C, Tatge H, Smith AM (1996) The elongation of amylose and amylopectin chains in isolated starch granules. Plant J 10:1135–1143

Devaux MF, de Monredon FLD, Guibert D, Novales B, Abecassis J (1998) Particle size distribution of break, sizing and middling wheat flours by laser diffraction. J Sci Food Agric 78:237–244

Dhaliwal L, Nanda G, Singh H, Dhaliwal H, Kaur H (1994) Inheritance of protein content and sds-sedimentation value in two crosses of bread wheat (*Triticum aestivum* L.). Crop Improvement 21:65–71

Feiz L, Martin J, Giroux M (2009) Creation and functional analysis of new Puroindoline alleles in *Triticum aestivum*. Theor Appl Genet 118:247–257

Giroux M, Morris C (1997) A glycine to serine change in puroindoline b is associated with wheat grain hardness and low levels of starch-surface friabilin. Theor Appl Genet 95:857–864

Giroux MJ, Morris CF (1998) Wheat grain hardness results from highly conserved mutations in the friabilin components puroindoline a and b. Proc Natl Acad Sci 95:6262–6266

Giroux MJ, Talbert L, Habernicht DK, Lanning S, Hemphill A, Martin JM (2000) Association of puroindoline sequence type and grain hardness in hard red spring wheat. Crop Sci 40:370–374

Gupta P, Kulwal P, Rustgi S (2005) Wheat cytogenetics in the genomics era and its relevance to breeding. Cytogenet Genome Res 109:315–327

Gupta R, Khan K, Macritchie F (1993) Biochemical basis of flour properties in bread wheats. I. Effects of variation in the quantity and size distribution of polymeric protein. J Cereal Sci 18:23–41

Gupta RB, Batey IL, MacRitchie F (1992) Relationship between protein composition and functional properties of wheat flours. Cereal Chem 69:125–131

Knott CA, Van Sanford DA, Souza EJ (2009) Genetic variation and the effectiveness of early-generation selection for soft winter wheat quality and gluten strength. Crop Sci 49:113–119

Kumar P, Yadava R, Gollen B, Kumar S, Verma R, Yadav S (2011) Nutritional contents and medicinal properties of wheat: A review. Life Sci Med Res 22:1–10

Law C, Young C, Brown J, Snape J, Worland A (1978) The study of grain protein control in wheat using whole chromosome substitution lines. Seed protein improvement by nuclear technique. Proceedings of the two research co-ordination meetings, Baden, 28 March–1 April 1977 and Vienna, 26–30 September 1977 Aneuploids in wheat protein improvement, pp 483–502

Li Z, Rahman S, Kosar-Hashemi B, Mouille G, Appels R, Morell M (1999) Cloning and characterization of a gene encoding wheat starch synthase I. Theor Appl Genet 98:1208–1216

Mattern PJ, Morris R, Schmidt JW, Johnson VA (1973) Locations of genes for kernel properties in the wheat variety "Cheyenne" using chromosome substitution lines. Nebraska Agricultural Experiment Station. In: Sears ER, Sears LMS (eds) Proceedings of the international wheat genetics symposium, Agricultural Experimental Station, Univeristy of Missouri, Columbia, 1–6 August 1973, pp 703–707

Miura H, Tanii S, Nakamura T, Watanabe N (1994) Genetic control of amylose content in wheat endosperm starch and differential effects of three Wx genes. Theor Appl Genet 89:276–280

Morris CF (2002) Puroindolines: the molecular genetic basis of wheat grain hardness. Plant Mol Biol 48:633–647

Morris R, Sears ER (1967) The cytogenetics of wheat and its relatives. In: Quisenberry KS & Reitz LP (eds) Wheat and wheat improvement, American Society of Agronomy, Madison, pp 19–87

Nakamura H (2000) The high-molecular-weight glutenin subunit composition of Japanese hexaploid wheat landraces. Crop Past Sci 51:673–677

Nakamura T, Vrinten P, Saito M, Konda M (2002) Rapid classification of partial waxy wheats using PCR-based markers. Genome 45:1150–1156

Naranjo T, Corredor E (2004) Clustering of centromeres precedes bivalent chromosome pairing of polyploid wheats. Trends Plant Sci 9:214–217

Payne PI, Nightingale MA, Krattiger AF, Holt LM (1987) The relationship between HMW glutenin subunit composition and the bread-making quality of British-grown wheat varieties. J Sci Food Agric 40:51–65

Rajaram S (2001) Prospects and promise of wheat breeding in the 21st century. Euphytica 119:3–15

Ram S, Boyko E, Giroux M, Gill B (2002) Null mutation in puroindoline a is prevalent in Indian wheats: puroindoline genes are located in the distal part of 5DS. J Plant Biochem Biotechnol 11:79–83

Riley R, Chapman V (1958) Genetic control of the cytologically diploid behaviour of hexaploid wheat. Nature 182:713–715

Rogers W, Sayers E, Ru K (2001) Deficiency of individual high molecular weight glutenin subunits affords flexibility in breeding strategies for bread-making quality in wheat *Triticum aestivum* L. Euphytica 117:99–109

Shewry P, Halford N, Tatham A (1992) High molecular weight subunits of wheat glutenin. J Cereal Sci 15:105–120

Sial M, Arain MA, Khanzada S, Naqvi MH, Dahot MU, Nizamani NA (2005) Yield and quality parameters of wheat genotypes as affected by sowing dates and high temperature stress. Pak J Bot 37:575

Singh R, Huerta-Espino J, Sharma R, Joshi A, Trethowan R (2007) High yielding spring bread wheat germplasm for global irrigated and rainfed production systems. Euphytica 157:351–363

Smith AM, Denyer K, Martin CR (1995) What controls the amount and structure of starch in storage organs? Plant Physiol 107:673

Snape J, Hyne V, Aitken K (1993) Targeting genes in wheat using marker-mediated approaches. In: Proceedings of the 8th international wheat genetics symposium, pp 749–759

Vrinten PL, Nakamura T (2000) Wheat granule-bound starch synthase I and II are encoded by separate genes that are expressed in different tissues. Plant Physiol 122:255–264

Wanjugi H, Hogg A, Martin J, Giroux M (2007) The role of puroindoline A and B individually and in combination on grain hardness and starch association. Crop Sci 47:67–76

Yan L, Fairclough R, Bhave M (1998) Molecular evidence supporting the origin of the B genome of Triticum turgidum from T. speltoides. In: Proceedings of 9th international wheat genetics symposium, pp 119–121

Zhao L, Zhang K-P, Liu B, Deng Z-y, Qu H-L, Tian J-C (2010) A comparison of grain protein content QTLs and flour protein content QTLs across environments in cultivated wheat. Euphytica 174:325-335

Chapter 2
Crop Competitiveness

Abstract Organic agriculture is being practiced in many countries around the globe and the area under organic agriculture is gradually increasing. Many biotic and abiotic factors affect wheat production in the field; however, competition for water, light, space and nutrients is often severe under organic systems due to the existence of weeds in a larger number. Weed management in organic systems is more challenging because the organic system prohibits use of herbicides. Moreover, weeds are rapidly becoming resistant to herbicides. This has led the scientific community to use other management strategies such as breeding cultivars with improved competitive ability to cope with weed's infestation. Breeding for crop competitive ability requires selection of traits that confer competiveness against various stresses, which is becoming a main objective in breeding cultivars for organically managed systems. Competitive ability traits reduce weed germination, growth, establishment and seed set, ultimately leading towards an improvement in grain yield. This Chapter reviews the importance of crop competiveness and highlights various morphological traits that confer competitive ability in crop plants with respect to developing new wheat cultivars for organically managed lands with enhanced competiveness.

Keywords Competitive ability · Crop morphology · Interference · Intraspecific competition · Organic agriculture · Weeds · *Triticum aestivum*

Wheat (*Triticum aestivum* L.) covers the largest area (217 m ha) under any single crop in the world (FAOSTAT 2012). It thus contributes towards the use of large amounts of synthetic fertilizers and pesticides. Agriculture has made tremendous progress to feed the ever-increasing world population but at the same time we are facing serious problems and challenges. This is especially true in terms of overuse and inappropriate use of agro-chemicals to make the soil fertile and to control weeds and diseases. This has resulted in contamination of water, loss of genetic variability, and deterioration of soil quality ultimately affecting the global ecosystem. Sustainable agriculture in the form of organic farming/agriculture has emerged as a new concept to address these challenges. Organic agriculture can be defined as a *production system that aims to promote and enhance agro-ecosystem health while*

M. Asif et al., *Managing and Breeding Wheat for Organic Systems*,
SpringerBriefs in Agriculture, DOI: 10.1007/978-3-319-05002-7_2,
© The Author(s) 2014

discouraging the use of off-farm inputs (Reid et al. 2009). Basically, organic agriculture focuses on improving soil fertility and reducing the use of external inputs by avoiding chemical/synthetic fertilizers, insecticides/pesticides, pharmaceuticals, and genetically modified organisms (GMO). The Codex Alimentarius Commission, an international food standard organization established in 1963 by the Food and Agriculture Organization (FAO) of the United Nations (UN) and the World Health Organization (WHO) defines organic agriculture in detail: *Organic agriculture is a holistic production management system which promotes and enhances agro-eco-system health including biodiversity, biological cycles and soil biological activity. It emphasizes the use of management practices in preference to the use of off-farm inputs, taking into account that regional conditions require locally adapted systems. This is accomplished by using where possible, agronomic, biological and mechanical methods, as opposed to using synthetic materials, to fulfill any specific function within the system* (Sligh and Christman 2003).

Organic agriculture is being practiced in many countries around the globe and the area under organic agriculture is expanding yearly. In 2011, organic agriculture was practiced on more than 37.2 m ha worldwide with 1.8 million producers and the global market reaches US$62.8 billion (Willer and Kilcher 2012). The total organic land constitutes 0.86 % of the total agricultural land (Willer and Kilcher 2012). The Australian/Oceania continent accounted for 12.2 m ha followed by Europe (10.6), Latin America (6.8), Asia (3.7), North America (2.8), and Africa (1.1) and shared 33, 29, 18, 10, 7, and 3 % of the total land area under organic production, respectively (Willer and Kilcher 2012). The demand for organic products is mainly present in Europe and North America where it generates 97 % of the total global revenues (Willer and Kilcher 2012). The North American demand is increasing at a rapid pace and will eventually overtake Europe to become the world's largest market (Reid et al. 2009).

2.1 Importance of Crop Competitiveness

The availability and use of chemical fertilizers, pesticides, and herbicides after the Green Revolution has resulted in a tremendous increase in grain yield of various crops, including wheat. Farmers have successfully grown crops in weed-free environments over the last five decades through the widespread use of herbicides. Many types of weeds infest wheat crop (Fig. 2.1) and large variability in losses due to weeds have also been reported, such as 31–41 % in rice (*Oryza Sativa*) (Bhatt and Tewari 2006), 16–29 % in barley (*Hordeum vulgare*) (Didon and Bostrom 2003), up to 40 % in peas (*Pisum sativum*), 40 % in canola (*Brassica* sp.) (Harker 2001), 77 % in sugar beet (*Beta vulgaris*), and 100 % in onions (*Allium cepa*) (Vanheemst 1985). Mason and Spaner (2006) reported a yield reduction of 40 % in trials conducted on organically managed land in Canadian environment using 32 spring bread wheat (*Triticum aestivum* L.) cultivars. The estimated crop losses in Canada due to weeds are $984 million in 58 different commodities, with a

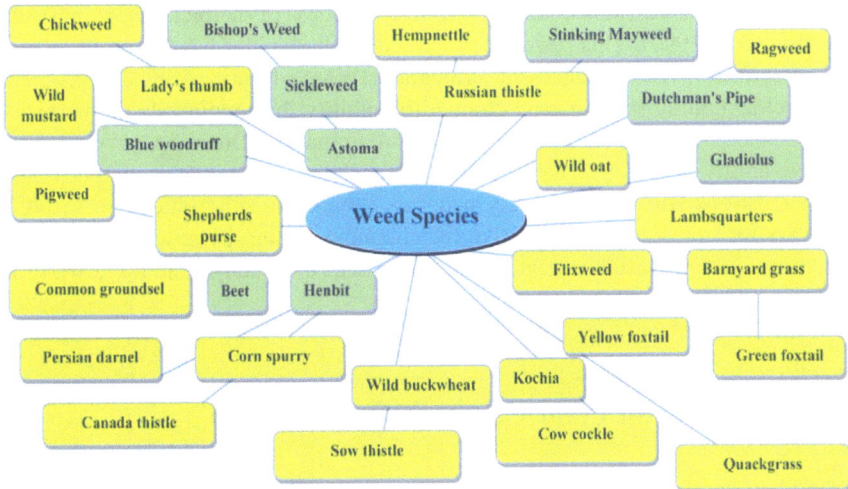

Fig. 2.1 Reported weed species (their common names) in Canada (*yellow*) and other parts (*green*) of the world that infest wheat field

share of $372 and $612 million of eastern and western Canada, respectively (Swanton et al. 1993). It has been reported that 1–3 Canada thistle plants/m^2 can reduce wheat yield by 12 % whereas a reduction of 36 % has been reported with 13–20 thistle plants/m^2 (Infanger 1956). Similarly, weed competition from *Avena sterilis* L. resulted in 23 and 19 % lower straw and grain yield in wheat, respectively. Gonzaliz-Ponce and Santin (2001) suggested that *A. sterilis* competes with wheat plants due to its emergence at the same time and great demand for nutrients and water. In another study, *A. fatua* and *Phalaris minor* were termed as "troublesome grassy" weeds that had shown a 30 % grain yield reduction in wheat. In addition to grain yield reduction, weed infestation also affects tillering ability, number of spikelets/spike, and quality (Khan et al. 2012). The bulblets of wild garlic and onion in wheat fields considerably influenced bread texture, color, and flavor and thus spoil flour quality (Qasem 2003). Moreover, increased herbicide resistances in weeds (Powles et al. 1997) have resulted in grain yield reductions in various crops, especially in cereals. This has directed scientists/researchers to take into account the importance of crop competiveness for weed management in the field. Semi-dwarf wheat cultivars, having height reducing genes (*Rht*) in the background, developed as a result of Green Revolution, are input responsive and high yielding but may be less competitive against weeds than older wheat cultivars developed before the Green Revolution (Mason and Spaner 2006).

The rate of weeds becoming resistant to herbicides is increasing at a very rapid pace due to continuous use of herbicides on a yearly basis (McDonald and Gill 2009). Deirdre and Donald (2001) reported various cases of herbicides resistance in 20 different wheat growing countries around the world. *Lolium rigidum*, *Avena* spp., *Raphanus raphanistrum*, along with some other weed species have shown most significant resistance against various groups of herbicides in Australia (Preston et al. 1999). This resistance of weeds to herbicides has forced farmers to use integrated weed management strategies (IWMS) to cope with this problem. The IWMS mainly rely on crop competitive ability that minimizes the herbicide cost/reliance and reduces environmental contamination. Breeding for crop competitive ability is basically the selection of those traits that confer competiveness against various stresses. This may become a main breeding objective: to breed cultivars for organically competitive management system by reducing weed germination, growth, establishment, and seed set, ultimately leading toward an improvement in grain yield.

2.2 What is Crop Plant Competition?

Crop plant competition (Fig. 2.2) is the demand for a common pool of resources (water, light, space, and nutrients) that is often limited in field conditions and is a major determinant of crop yield, especially in case of cereals. Competition can be of two types; intraspecific (between plants of the same species) and interspecific (between plants of different species). In organic farming, the interspecific competition is severer than conventional management systems due to the existence of more weeds. Christensen (1993) argued that interaction between plants is "interference" that can be positive, negative, or neutral. In organic environments, negative interference always exists between crop plants and weeds and is referred to as competition. Various studies have been conducted to find out the relative competitiveness of various crops against weeds (Holman et al. 2004; Lutman et al. 1994; Pavlychenko and Harrington 1934). The competitive ability of these crop species in order of their competitiveness is shown in Fig. 2.3. Weed–crop competition or competitive ability of a crop can be estimated in two ways, i.e., the suppression of weed growth by crop plants or grain yield reduction of crop plants by weeds. This dominant–recessive relationship of crop and weed plants, or vice versa, depends on genetic, management, and environmental factors. Challaiah et al. (1986) and Lemerle et al. (1996a) emphasized that crop tolerance and weed suppression are two separate characteristics, not necessarily present in the same genotype but both are highly correlated to each other (especially in wheat). The ideal genotype should have these two traits and therefore can perform equally in organic and conventional management systems by tolerating as well as suppressing weeds. Weed–crop competition can be divided into three basic categories; (1) weed-related factors including weed species, weed density, weed emergence time, duration of weed presence in the field, (2) crop-related factors including genetic response, crop species, crop cultivars, canopy cover, plant height, light interception rate, leaf width and angle, and (3) management-related factors including seeding rate, time, and density, time of irrigation and fertilization, soil and other environmental conditions, and efficiency of weed control methods. The most important factors of competitiveness are discussed in detail.

Fig. 2.2 Schematic diagram of crop competition with biotic (weeds and diseases) and abiotic factors

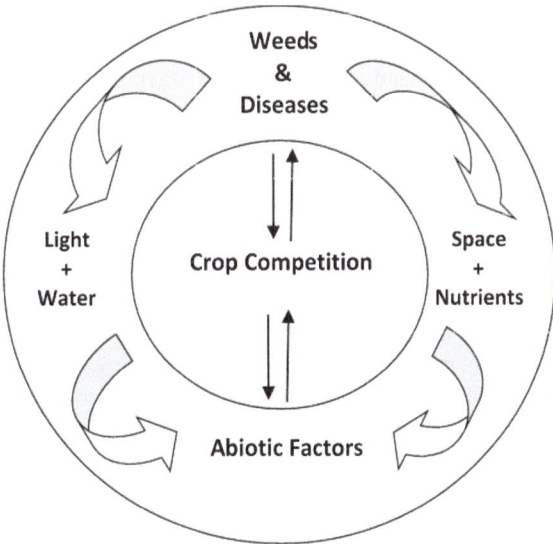

Fig. 2.3 Order of competitiveness in various crops against weeds

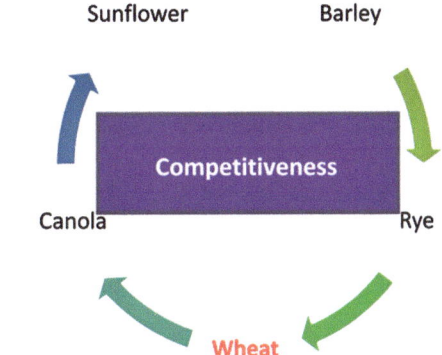

2.3 Association of Plant Traits to Competitiveness

Various plant traits affect competitive ability and vice versa. The negative effects of plant competition include shortage of water, nutrients and light, susceptibility to diseases and lodging, biomass/yield reduction, etc. Salisbury (1936) wrote that *competition is severe in the plant world, no one can deny and this is operative in all phases of development.* Therefore, it is difficult to find a crop plant in field conditions that have not been affected by its neighboring community, either of the same or different species. The existence of competition in the field makes it difficult to determine plant traits that are contributing to plant competitive ability because they are not only affected by the management practices (such as seeding rate, plant spacing, sowing depth) but also due to the environment and genotype x environment x management interactions. Moreover, competition is a population phenomenon and not simply the interaction with a single crop plant. Several authors advocate the use of morphological, physiological, biochemical, or behavioral characteristics to measure competitive ability (Lemerle et al. 2001a, b). However, any trait can confer competitiveness if it has the ability to rapidly utilize available resources and can increase its above-ground biomass as compared to its neighboring plants with which it competes (Donald 1968). Crop competitiveness is a genetic trait that is influenced by genotype, management, environment (their interactions), and is dependent on traits like plant height, early season vigor, light interception, leaf area index (LAI) and architecture, tillering ability, canopy structure, crop ground cover, earliness, nutrient use efficiency, lodging, and disease resistance. Mason and Spaner (2006) suggested that plant characteristics that exhibit a high degree of crop competitiveness against weeds are highly desirable in organically managed lands. Moreover, identification of new traits that confer competitiveness against weeds is a perquisite to breed/develop new cultivars for organic agriculture. Crop competitiveness cannot be attributed to a single plant trait; rather it is a sum of interaction of several plant characteristics. Early season vigor is directly linked with the competitive ability of the crop (Bertholdsson 2005) and later in the crop growing season, it confers competition against weeds (Huel and Hucl 1996). Plant traits that exhibit a positive correlation with early season vigor include rapid early growth rate (Froud-Williams 1999), LAI (Huel and Hucl 1996; Lemerle et al. 1996a, b) and tillering ability (Lemerle et al. 2001a), but it is also influenced by management practices like seed rate/plant density, seeding depth, and row spacing.

The relative emergence time of crop plants strongly influences competitive ability in cereals. If emergence of crop plants occurs before weeds, they will be able to use much of the limited resources as compared to weed plants which will ultimately give rise to a competitive advantage. Relative emergence time was studied between barley and *Sonchus arvensis* (sow thistle), where the authors reported that emergence of crop plants 4 days before weeds resulted in 90 % of the above-ground crop biomass. An opposite scenario occurred when crop plants emerged 8 and 26 days after weed plants and this late emergence resulted in a decreased above-ground crop biomass coverage by 50 and 10 %, respectively (Eckersten et al. 2008, 2010).

Early season vigor is one of the most important traits contributing to crop competitiveness. A plant that efficiently utilizes resources during early life cycle/growth stages in terms of greater photosynthetic active radiation (PAR), accumulation of biomass, and tillering improves competitive ability. Early season vigor can be easily determined on the basis of early seedling size (width and length of seedling leaves), germination percentage (Lemerle et al. 2001a, b), and early plant biomass (Zerner et al. 2008). In a durum wheat experiment, cultivars possessing high early biomass accumulation and greater PAR were found best competitors against weeds (Lemerle et al. 1996b). This study also reported a strong negative correlation between first leaf length and width, and seedling biomass accumulation with grain yield loss and weed dry matter. Similar results have been reported in barley, oats, and wheat, where early season vigor was found to be strongly correlated with first leaf length and width (Lopez et al. 1996). Early season vigor is highly dependent on relative growth rate (increase in plant biomass per unit of time) of a crop. High relative growth rate in early growth staged confers better competitive ability (Grime 1979).

Plant height also has a pronounced effect on grain yield and is one of the most extensively reported and desired traits with respect to crop competitiveness. Mason et al. (2007), Cousens et al. (2003a, b) and Gooding et al. (1993) reported that plant height is directly linked to plant competitiveness. This relationship was supported previously by Cudney et al. (1991) who mentioned that taller cultivars are more competitive than shorter ones because of better light interception, and a direct association with photosynthetic activity of the plant in general. The importance of *Rht* genes in wheat was evaluated with respect to short-statured plants in an organic environment where the authors reported high weed (*Alopecurus myosuroides*) infestation and linked it to a reduction in shading ability of short-statured wheat plants (Gooding et al. 1997). This reduced shading ability caused increased penetration of photosynthetically active radiation through the wheat canopy. Wheat was reported to be the weakest competitor among cereals (Lemerle et al. 1995). With the introduction of semi-dwarf wheat cultivars during the Green Revolution, *Rht* (height reducing) genes are being incorporated and these cultivars are not performing well in organically managed systems due to less suppression of weeds, particularly as a result of their reduced plant height.

Thus, breeding for enhancing grain yield in wheat has decreased competitive ability to a considerable degree that has resulted in reducing early leaf area development (Rebetzke and Richards 1999) which has been considered as one of the main traits contributing toward competitive ability (Coleman et al. 2001). Similar findings have also been reported by Watson et al. (2006) in an experimental study involving 29 barley cultivars under Canadian environment where authors reported that if cultivars do not exhibit enough plant height with respect to their competitors, they must have enhanced and elevated leaves well above the ground and covered more surface area to maximize their light interception. Tall wheat cultivars were found to be capable of using more PAR and solar radiation in the range of 400–700 nm, and thus helped wheat suppress weeds (Champion et al. 1998). Tall rice cultivars also possess greater competitiveness and it was observed that plant height at very early stage of plant development can be a good indicator of competitive ability and to screen lines in standard nurseries (Caton et al. 2003; Moukoumbi et al. 2011). Plant height was found to have positive and significant correlation with competitive ability of crops; and the shortest cultivars when grown in a weedy competitive environment resulted in greater yield reductions and increased weed biomass (Huel and Hucl 1996).

Weed population is directly dependent on plant height of crop plants and weed species. It is also a major determinant of crop competitive ability because competitiveness of a given crop varies between different weed species. In an experiment on wheat competitive ability using near-isogenic lines, plant height showed a linear relationship to decrease weed seed production (Seefeldt et al. 1999). A similar trend of plant height in 20 winter wheat cultivars was reported by Wicks et al. (1986) who mentioned that cultivars taller than 83 cm reduced weed growth to a greater extent by capturing more light than shorter cultivars. In a similar study, Yenish and Young (2004) reported that tall (~ 130 cm) wheat cultivars decreased goat grass (*Aegilops cylindrica*) by 45 % as compared to short-statured (~ 100 cm) cultivars However, two shorter cultivars were also termed the best competitors with weeds suggesting that there are traits conferring competitive ability other than plant height (Wicks et al. 1986).

Crop canopy development, LAI and leaf shape/structure, angle, and number are other traits that are linked to competitiveness. It has been reported that semi-dwarf cultivars of wheat have smaller cells than older/conventional cultivars and this shorter cell size contributes to reduced leaf area and shorter coleoptiles length (Vandeleur and Gill 2004). The opposite or no differences were reported in another study conducted by Entz et al. (1992) suggesting that these undesirable traits can be improved by breeding (Ellis et al. 2004). Jennings and Aquino (1968) suggested that elevated competitive ability of rice was directly linked with high LAI. Strongly competitive plants have strong and spreading leaves. In an interesting study conducted on two wheat cultivars (Spark and Avalon), Seavers and Wright (1999) found Avalon to be a strong competitor due to the presence of re-curved leaves and more rapid canopy development than the more upright growing Spark. Gaudet and Keddy (1988) determined crop competitive ability and reported a linear relationship between easily observable characteristics and competitiveness ($r = 0.74$), where plant biomass explained 63 % of the total variation. Other characteristics, including leaf shape, height, canopy area, and diameter explained residual variation. Leaf area index was negatively correlated with weed seed yield and had no impact on grain yield reduction in wheat (Huel and Hucl 1996). In later growth stages, flag leaf characteristics like length, width, and angle also played a key role in suppressing weeds and maintaining a healthy crop stand. A strong negative correlation between flag leaf length and grain yield reduction was reported in the same study. However, care must be taken if breeding is done for dryland or drought prone areas because a high LAI can have a negative impact on grain yield (Richards 1983) under water limited conditions. Richardson et al. (2001) proposed that breeding for organic agriculture is more practicable if screening for candidate lines is done on the basis of visual observation for plant growth habit, leaf habit, leaf angle, leaf inclination, and visual scoring or image analysis of canopy coverage, than from light interception or a quantification of weed biomass per unit area.

Light interception is another important trait linked with weed suppression or with a resistance to yield suppression by weeds. It has been well documented that cultivars of winter wheat and spring cereals vary considerably more for shoot competitive ability than root competition, suggesting the importance of light interception in crop competition (Lucas and Froud 1994; Satorre and Snaydon 1992). Shoot morphological traits, including long main shoot during tillering, small leaf angle, and less time of emergence are important traits directly associated with competition for light (Didon 2002). Moreover, wheat cultivars with more ground cover exhibit high light interception. Ground cover is directly affected by leaf inclination, plant height, and row spacing. Later developing or smaller weeds are more susceptible to light competition and, therefore, it was recommended that selection for taller planophile wheat cultivars with narrow row spacing (12–17 cm) may provide an effective strategy to control losses due to weeds as compared to erectophile genotypes (Drews et al. 2004).

Overall, canopy development of a specific crop is highly dependent on its light interception rate, especially during the early stages of its life cycle. Fischer et al. (2001) described the direct association of light interception with crop competitive ability due to a negative correlation of LAI and PAR with weed biomass in rice. Greater leaf area and weight were present in the most competitive rice cultivar (Johnson et al. 1998) and these two traits reduced the amount of photosynthesis that is required to produce a given leaf area for light interception (Dingkuhn et al. 1999). Light interception varied considerably from one genotype to another and was also highly dependent on management practices like row spacing, fertilizer application, and number of irrigations during crop growing season. Light interception increases continuously until 90 days after sowing in wheat and then decreases slowly until maturity (Mishra et al. 2009). These results suggest that selection for light interception during early growth stages could improve competitiveness by increasing PAR interception and radiation use efficiency (RUE) (Zhou et al. 2011).

High tillering ability in cereals is also a desirable trait to obtain maximum grain yield per unit area and has been reportedly linked to above-ground canopy coverage and shading ability (Didon and Hansson 2002; Korres and Froud-Williams 2002). In an experiment to study the differences of competitive ability between old and new wheat cultivars, Fang et al. (2011) reported superior competitiveness of an old wheat cultivar/land race (Pinglang 40) over a modern cultivar (Changwu 135) due to greater tillering capacity, taller plants, larger root system, and LAI of the Pinglang 40.

References

Bertholdsson NO (2005) Early vigour and allelopathy—two useful traits for enhanced barley and wheat competitiveness against weeds. Weed Res 45:94–102

Bhatt MD, Tewari A (2006) Losses in growth and yield attributes due to weed composition in transplanted paddy in terai region. Scientific World 4:99–101

Caton BP, Cope AE, Mortimer M (2003) Growth traits of diverse rice cultivars under severe competition: implications for screening for competitiveness. Field Crops Res 83:157–172

Challaiah O, Ramsel RE, Wicks GA, Burnside OC, Johnson VA (1986) Evaluation of the weed competitive ability of winter wheat cultivars. Proceedings of the North Central Weed Control Conference Tasmanian Weeds Society, Hobart, Tasmania, Australia, pp 85–91

Champion GT, Froud-Williams RJ, Holland JM (1998) Interactions between wheat (*Triticum aestivum* L.) cultivar, row spacing and density and the effect on weed suppression and crop yield. Annals Appl Biol 133:443–453

Christensen S (1993) Weed Suppression in Cereal Varieties. Min. Agric. Statens Planeavlsforsog, Denmark, p 104

Coleman RD, Gill GS, Rebetzke GJ (2001) Identification of quantitative trait loci for traits conferring weed competitiveness in wheat (*Triticum aestivum* L.). Aust J Agric Res 52:1235–1246

Cousens RD, Barnett AG, Barry GC (2003a) Dynamics of competition between wheat and oat: I. Effects of changing the timing of phenological events. Agronomy J 95:1295–1304

Cousens RD, Rebetzke GJ, Barnett AG (2003b) Dynamics of competition between wheat and oat: II. Effects of dwarfing genes. Agronomy J 95:1305–1313

Cudney DW, Jordan LS, Hall AE (1991) Effect of wild oat (*Avena fatua*) infestations on light interception and growth-rate of wheat (*Triticum aestivum*). Weed Sci 39:175–179

Deirdre L, Donald T (2001) World wheat and herbicide resistance. Herbicide Resistance and World Grains. CRC Press, FL

Didon UME (2002) Variation between barley cultivars in early response to weed competition. J Agron Crop Sci 188:176–184

Didon UME, Bostrom U (2003) Growth and development of six barley (*Hordeum vulgare* ssp *vulgare* L.) cultivars in response to a model weed (*Sinapis alba* L.). J Agron Crop Sci 189:409–417

Didon UME, Hansson ML (2002) Competition between six spring barley (*Hordeum vulgare* ssp *vulgare* L.) cultivars and two weed flora in relation to interception of photosynthetic active radiation. Biol Agric Hortic 20:257–274

Dingkuhn M, Johnson DE, Sow A, Audebert AY (1999) Relationships between upland rice canopy characteristics and weed competitiveness. Field Crops Res 61:79–95

Donald CM (1968) Breeding of crop ideotypes. Euphytica 17:385–403

Drews S, Juroszek P, Neuhoff D, Kopke U (2004) Optimizing shading ability of winter wheat stands as a method of weed control. Z Pflanzenk Pflanzens-J Plant Dis Prot 19:545–552

Eckersten H, Andersson L, Holstein F, Mannerstedt Fogelfors B, Lewan E, Sigvald R, Torssell B, Karlsson S (2008) An evaluation of climate change effects on crop production in Sweden. Swedish University of Agricultural Sciences, Department of Crop Production Ecology, Report No 6, ISBN 978-91-576-7237-7, Uppsala, Sweden

Eckersten H, Lundkvist A, Torssell B (2010) Comparison of monocultures of perennial sow-thistle and spring barley in estimated shoot radiation-use and nitrogen-uptake efficiencies. Acta Agr Scand Section B-Soil Plant Sci 60:126–135

Ellis MH, Rebetzke GJ, Chandler P, Bonnett D, Spielmeyer W, Richards RA (2004) The effect of different height reducing genes on the early growth of wheat. Funct Plant Biol 31:583–589

Entz MH, Gross KG, Fowler DB (1992) Root-growth and soil-water extraction by winter and spring wheat. Can J Plant Sci 72:1109–1120

Fang Y, Liu L, Xu BC, Li FM (2011) The relationship between competitive ability and yield stability in an old and a modern winter wheat cultivar. Plant Soil 347:7–23

FAOSTAT (2012) http://faostat.fao.org

Fischer AJ, Ramirez HV, Gibson KD, Pinheiro BD (2001) Competitiveness of semidwarf upland rice cultivars against palisadegrass (*Brachiaria brizantha*) and signalgrass (*B. decumbens*). Agronomy J 93:967–973

Froud-Williams RJ (1999) A biological framework for developing a weed management support system for weed control in winter wheat: weed seed biology. In: Proceedings Brighton conference weeds, pp 747–752

Gaudet CL, Keddy PA (1988) A comparative approach to predicting competitive ability from plant traits. Nature 334:242–243

Gonzaliz-Ponce R, Santin I (2001) Competitive ability of wheat cultivars with wildoats depending on nitrogen fertilization. Agronomie 21:119–125

Gooding MJ, Cosser ND, Thompson AJ, Davies WP, Froud-Williams RJ (1997) The effect of cultivar and *Rht* genes on the competitive ability, yield and bread-making qualities of organically grown wheat. Proceedings of the 3rd ENOF workshop: resource use in organic farming, Ancona, pp 113–121

Gooding MJ, Thompson AJ, Davies WP (1993) Interception of photosynthetically active radiation, competitive ability and yield of organically grown wheat varieties. Aspects of applied biology/Association of Applied Biologists 34:355–362

Grime JP (1979) Plant strategy and vegetation processes. Wiley, Chichester, UK

Harker KN (2001) Survey of yield losses due to weeds in central Alberta. Can J Plant Sci 81:339–342

Holman JD, Bussan AJ, Maxwell BD, Miller PR, Mickelson JA (2004) Spring wheat, canola, and sunflower response to Persian darnel (*Lolium persicam*) interference. Weed Technol 18:509–520

Huel DG, Hucl P (1996) Genotypic variation for competitive ability in spring wheat. Plant Breeding 115:325–329

Infanger CA (1956) Economic consequences of 2,4-D in controlling Canada thistle in irrigated spring wheat. Montana State College

Jennings PR, Aquino RC (1968) Studies on competition in rice. III. The mechanism of competition among phenotypes. Evolution 22:529–542

Johnson DE, Dingkuhn M, Jones MP, Mahamane MC (1998) The influence of rice plant type on the effect of weed competition on *Oryza sativa* and *Oryza glaberrima*. Weed Res 38:207–216

Khan IA, Hassan G, Khan SA, Shah SMA (2012) Wheat-wild oats interactions at varying densities and proportions. Pak J Bot 44:1053–1057

Korres NE, Froud-Williams RJ (2002) Effects of winter wheat cultivars and seed rate on the biological characteristics of naturally occurring weed flora. Weed Res 42:417–428

Lemerle D, Gill GS, Murphy CE, Walker SR, Cousens RD, Mokhtari S, Peltzer SJ, Coleman R, Luckett DJ (2001a) Genetic improvement and agronomy for enhanced wheat competitiveness with weeds. Aust J Agric Res 52:527–548

Lemerle D, Verbeek B, Coombes N (1995) Losses in grain yield of winter crops from *Lolium rigidum* competition depend on crop species, cultivar and season. Weed Res 35:503–509

Lemerle D, Verbeek B, Cousens RD, Coombes NE (1996a) The potential for selecting wheat varieties strongly competitive against weeds. Weed Res 36:505–513

Lemerle D, Verbeek B, Martin P (1996b) Breeding wheat cultivars more competitive against weeds

Lemerle D, Verbeek B, Orchard B (2001b) Ranking the ability of wheat varieties to compete with *Lolium rigidum*. Weed Res 41:197–209

Lopez C, Richards RA, Farquhar GD, Williamson RE (1996) Seed and seedling characteristics contributing to variation in early vigor among temperate cereals. Crop Sci 36:1257–1266

Lucas BC, Froud WRJ (1994) The role of varietal selection for enhanced crop competitiveness in winter wheat. Aspects Appl Biol 40:343–350

Lutman PJW, Dixon FL, Risiott R (1994) The response of four spring-sown combinable arable crops to weed competition. Weed Res 34:137–146

Mason HE, Navabi A, Frick BL, O'Donovan JT, Spaner DM (2007) The weed-competitive ability of Canada western red spring wheat cultivars grown under organic management. Crop Sci 47:1167–1176

Mason HE, Spaner D (2006) Competitive ability of wheat in conventional and organic management systems: A review of the literature. Can J Plant Sci 86:333–343

McDonald GK, Gill GS (2009) Improving crop competitiveness with weeds: Adaptations and trade-offs. Crop Physiology. Academic Press, San Diego, pp 449–488

Mishra AK, Tripathi P, Pal RK, Mishra SR (2009) Light interception and radiation use efficiency of wheat varieties as influenced by number of irrigations. J Agrometeorol 11:140–143

Moukoumbi YD, Sie M, Vodouhe R, Bonou W, Toulou B, Ahanchede A (2011) Screening of rice varieties for their weed competitiveness. Afr J Agric Res 6:5446–5456

Pavlychenko TK, Harrington JB (1934) Competitive efficiency of weeds and cereal crops. Can J Res 10:77–94

Powles SB, Preston C, Bryan IB, Jutsum AR (1997) Herbicide resistance: Impact and management. In: Sparks DL (ed) Advances in agronomy, vol 58, pp 57–93

Preston C, Roush RT, Powles SB (1999) Herbicide resistance in weeds of southern Australia: why are we the worst in the world? Proceedings of the 12th Australian Weeds conference. Tasmanian Weeds Society, Hobart, Tasmania, Australia, pp 454–459

Qasem JR (2003) Weeds and their control. University of Jordan Publications Amman, Jordan 628

Rebetzke GJ, Richards RA (1999) Genetic improvement of early vigour in wheat. Aust J Agric Res 50:291–301

Reid T, Yang R-C, Salmon D, Spaner D (2009) Should spring wheat breeding for organically managed systems be conducted on organically managed land? Euphytica 169:239–252

Richards RA (1983) Manipulation of leaf-area and its effect on grain-yield in droughted wheat. Aust J Agric Res 34:23–31

Richardson AE, Hadobas PA, Hayes JE (2001) Extracellular secretion of *Aspergillus phytase* from Arabidopsis roots enables plants to obtain phosphorus from phytate. Plant J 25:641–649

Salisbury EJ (1936) Natural selection and competition. Proc Roy Soc Lond B 121:47–49

Satorre EH, Snaydon RW (1992) A comparison of root and shoot competition between spring cereals and *Avena fatua* L. Weed Res 32:45–55

Seavers GP, Wright KJ (1999) Crop canopy development and structure influence weed suppression. Weed Res 39:319–328

Seefeldt SS, Ogg AG, Hou YS (1999) Near-isogenic lines for *Triticum aestivum* height and crop competitiveness. Weed Sci 47:316–320

Sligh M, Christman C (2003) Who owns organic? The global status, prospects and challenges of a changing organic market. Rural Advancement Foundation International (RAFI), USA

Swanton CJ, Harker KN, Anderson RL (1993) Crop losses due to weeds in canada. Weed Technol 7:537–542

Vandeleur RK, Gill GS (2004) The impact of plant breeding on the grain yield and competitive ability of wheat in Australia. Aust J Agric Res 55:855–861

Vanheemst HDJ (1985) The influence of weed competition on crop yield. Agric Syst 18:81–93

Watson PR, Derksen DA, Van Acker RC (2006) The ability of 29 barley cultivars to compete and withstand competition. Weed Sci 54:783–792

Wicks GA, Ramsel RE, Nordquist PT, Schmidt JW, Challaiah (1986) Impact of wheat cultivars on establishment and suppression of summer annual weeds. Agronomy J 78:59–62

Willer H, Kilcher L (2012) The World of Organic Agriculture—Statistics and Emerging Trends Research Institute of Organic Agriculture (FiBL), Frick, and International Federation of Organic Agriculture Movements (IFOAM), Bonn

Yenish JR, Young FL (2004) Winter wheat competition against jointed goatgrass (*Aegilops cylindrica*) as influenced by wheat plant height, seeding rate, and seed size. Weed Sci 52:996–1001

Zerner MC, Gill GS, Vandeleur RK (2008) Effect of height on the competitive ability of wheat with oats. Agronomy J 100:1729–1734

Zhou XB, Chen YH, Ouyang Z (2011) Row spacing effect on leaf area development, light interception, crop growth and grain yield of summer soybean crops in Northern China. Afr J Agric Res 6:1430–1437

Chapter 3
Strategies to Enhance Competitive Ability

Abstract Weeds compete for limited resources under field conditions and cause quantitative and qualitative grain yield losses. Weeds cause losses in both conventional and organically managed lands but competition from weeds is more challenging to organic farmers as synthetic herbicides cannot be used. Various strategies have been proposed that can enhance competitive ability of crop plant by reducing weed seed production and crop grain yield losses. In this Chapter, various management strategies have been discussed in detail that can facilitate farming community to enhance competitive ability in different environments. Competitive ability is a genetic trait that varies among crop cultivars. Research on wheat competitive ability has been conducted in various parts of the world suggesting that genetic differences exist for competitive ability between crop species and also within cultivars. This variation for competitive ability can be used to breed cultivars for improved competitiveness. Recent advancements in the field of biotechnology, especially relating to DNA marker technology have broadened the prospects of selecting suitable/desirable genotypes based on DNA markers. The chapter also highlights various studies designed to identify genomic regions (QTL) conferring competitive ability in various cereal crops.

Keywords Biological control · Crop density · Cover crops · Crop rotation · Molecular markers · Quantitative trait loci

3.1 Management

Weeds compete for limited resources under field conditions and cause quantitative and qualitative grain yield losses. Weeds cause losses in both conventional and organically managed lands but competition from weeds is more challenging to organic farmers as synthetic herbicides cannot be used. The main objective of weed control management strategies is to reduce competition from weeds. Various biological, chemical, mechanical, cultural, and integrated weed management (IWM) strategies are employed to minimize weed infestations and decrease yield

M. Asif et al., *Managing and Breeding Wheat for Organic Systems*,
SpringerBriefs in Agriculture, DOI: 10.1007/978-3-319-05002-7_3,
© The Author(s) 2014

losses. Use of a specific weed control method depends on the cropping system, area, economic status of farmer, climatic and environmental conditions.

Preventive weed control measures are implemented either through governmental laws or through farmer's precautions. Each country has its own weed seed laws which regulates and identifies the movement of crop seed with a specific percentage of noxious weed seeds. For example, the Canada Seeds Act (CSA) of 1987 permits a specific amount of weed seeds to be present in crop seed and this specific amount varies depending on the crop. These standards are applicable to lentils (*Lens culinaris*), rye (*Lolium* sp.), buckwheat (*Fagopyrum esculentum*), barley (*Hordeum vulgare*), with slight verifications to wheat (*Triticum aestivum*), oat (*Avena sativa*), canola (*Brassica* sp.), and flax (*Linum usitatissimum*) as well. Similarly, the United States of America passed their Noxious Weed Control Act (NWCA) in 2004 as an amendment to the Federal Plant Protection Act (FPPA) of 2000. Weed eradication is the easiest method to eliminate all weed plants, weed seeds, and weed parts in the field conditions; however, complete elimination is impossible. It is easy to remove weed plants but extremely difficult to remove weed seeds and other vegetative and reproductive parts of weeds in the soil.

3.1.1 Biological Control

Biological control involves the suppression of weed growth and development by means of living agents such as fungi, bacteria, and insects. The use of grazing animals like sheep, goats, horses, and cattle have also been used to minimize weeds. Muller et al. (2011) reported the use of rust pathogens to control *Cirsium arvense* (Canada thistle). The fungus, *Phoma macrostoma* has proven highly effective to control broad leaf weeds and is undergoing registration in both the USA and Canada (Bailey 2010). Similarly, bacterium *Xanthomonas campestris* pv. p*oae* has shown promising results to control *Poaannua* L. (Imaizumi et al. 1997). Another successful example of biological weed control is by the use of *Chrysolina hyperici* (leaf beetle) to control *Hypericum perforatum* L. which is attributed to the fungus *Colletotrichum gloeosporioides* (Harris et al. 1969; Morrison et al. 1998). However, for long-term successful weed control, it is necessary that small number of weed hosts must be present in the field to ensure the quick elimination of weeds during early stages of their life cycle. Moreover, biological weed control involves the control of particular weed species by specific hosts/pathogens that limit its scope.

3.1.2 Cultural Control

Competitive ability also involves the establishment of a vigorous crop by cultural control methods such as crop density, seed rate, proper fertilization, seeding method, planting time, crop rotation, varietal selection, row spacing, and planting pattern.

3.1.2.1 Crop Density

It is well documented in the literature that weed abundance and weed biomass decreases with an increase in crop density, and vice versa. The relative advantages of varying seed rates have been appreciated for a long time where "thick" wheat planting (80 kg ha^{-1}) was recommended to obtain maximum grain yield for late sown or sparsely tillering cultivars (Downing 1921). An increase in crop density reduces the availability of resources (water and other nutrients) to the weeds and thus suppresses their growth and development. A significant weed suppression was reported by an increase in crop density from 350 to 800 plant m^{-2} in various parts of the world including Canada (Kirkland 1993), USA (Hashem et al. 1998), Denmark (Doll et al. 1995), and the United Kingdom (Korres and Froud-Williams 1997). Similar benefits of optimum crop density have also been reported in another study to suppress wild oat infestation in wheat. Six plants of wild oats m^{-2} reduced grain yield by 20 % in wheat with a cropping density of 100 plants m^{-2}; whereas in a cropping density of 700 plants m^{-2}, 38 wild oat plants were required to cause 20 % grain yield losses (Carlson and Hill 1985). Lemerle et al. (1996a) provided evidence of substantial biomass reduction in *Lolium rigidum* by an increase in wheat planting density from 150 to 200 plant m^{-2}. A similar trend was observed with respect to seed production in *L. rigidum* by increasing wheat seeding rate from 50 to 200 kg ha^{-1} (Peltzer 1999). The seeding rate of 200 kg ha^{-1} gave the highest grain yields in barley and wheat and decreased seed production 1998 in *Phalaris paradoxa* and *Avena* sp. (Radford et al. 1980; Walker et al. 1998). A seeding rate of 270 kg ha^{-1} produced higher yield in spring wheat competing with foxtail than rates of 130 or 70 kg ha^{-1} (Khan et al. 1996). A 10 % increase in the grain yield of winter wheat competing with goat grass was reported using seeding rate of 60 rather than 40 seeds m-^{2} (Yenish and Young 2004). An increase in wheat seeding rate from 75 to 150 kg ha^{-1} in Alberta, Canada decreased biomass 18 % and lowered the soil seed bank of wild oat by 46 % in the absence of herbicide application. This increased seeding rate enhanced wheat grain yield by 19 % and net economic return by 16 % (O'Donovan et al. 2006). Gosling et al. (2006) reported an exponential relationship of *Avena sterilis* (wild oat) to decreased barley yield. They reported a 10 % reduction in barley yield with wild oat densities of 20–80 panicles m^{-2}, but grain yield reduction was as high as 50 % when weed density was more than 300 panicles m^{-2}. Similar results with respect to grain yield reduction in barley have been reported with increasing the number of *Lolium rigidum* (swiss ryegrass) from 16 to 125 plants m^{-2} (Paynter and Hills 2009). Similarly, *Lamiuma plexicaule* (henbit) did not cause any grain yield reduction in wheat when its density was 18 plants m^{-2}; however, an increase in density from 18 to 82 and 155 plants m^{-2} decreased grain yield up to 13 and 38 % respectively (Cook 2007).

   In cereals, it is a well-established fact that grain yield can be improved by increasing seeding rate, but up to a certain level beyond which it remains constant/ unchanged or starts decreasing. Therefore, care must be taken to avoid any intraspecific competitions among crop plants. Moreover, higher seeding rate are also associated with extra seed costs for planting and adoption for higher seeding rate is variable among farming communities. Nevertheless, there is strong evidence to support the advantages of seeding rate higher than 100 kg ha^{-1} with respect to competitive ability, reducing weed densities, and increased net gains (Lemerle et al. 2001a).

3.1.2.2 Fertilization

The interaction between crop plants and weeds is greatly affected by fertility management during the cropping season. In organic management systems, crop plants rely on organic fertilizers or amendments that discharge nutrients, especially N, at slower rates than synthetic fertilizers (Magdoff 1995). The release of nutrients by fertilizers is dependent on soil properties, water availability, environmental conditions, and placement method. Uptake of nutrients by weeds and crop plants can also vary considerably and is dependent on source-sink ratio as well as types of weeds, crop cultivars, and time of fertilizer application. Rapid release of nutrients by synthetic fertilizers favor weed species because they are capable to uptake and use nutrients quickly during early growth stages of their life cycle (Liebman and Davis 2000). Therefore, slower release of nutrients on organically managed lands did not result in high crop-weed competition during early stages (Liebman and Davis 2000). However, it may result in the late emergence of weeds that will contribute to increase weed seed in the field and can increase weed populations in subsequent years (Barberi 2002). The uptake of nutrients (N, P, K) promotes stem elongation, leaf area, branching, and canopy development. Weeds are generally considered much more efficient at capturing these nutrients than crops (Alkamper 1976) and as a result; more green surface area was reported in weed (*Brassica hirta*) plants in a study where ammonium sulfate was applied to a mixture of barley and *Brassica hirta* (Liebman and Robichaux 1990). In a similar study, N application increased panicle production in *Avenua fatua* by 140 % and lowered grain yield in wheat by 49 % when *A. fatua* was present in high densities (Carlson and Hill 1985). The placement of fertilizers (organic) in bands very close to crop rows in organic systems can be done to maximize nutrient centration in the crop root zone that may help decrease competition between crop plants and weeds for nutrients. Fertilizer banding has shown promising results to increase crop yield and decrease weed biomass and density (Ditomaso 1995; Rasmussen et al. 1996). A comparison of wheat grown under organic and conventional management systems is presented in Figs. 3.1–3.7.

3.1.2.3 Row Spacing and Planting Pattern

Varying row width is another strategy to control weed populations. Decreasing in row spacing often favors crop plants with respect to competition against weeds. Narrow row spacing of crop plants (i.e., 20 cm) reduced weed growth by 37 % in

Fig. 3.1 Early stage wheat in conventional management system

Fig. 3.2 Early stage wheat in organic management system

sorghum (Wiese et al. 1964) and 55 % in peanuts (Buchanan and Hauser 1980) as compared to 30 cm. Decreasing row spacing from 78 to 38 cm along with the use early maturing corn hybrids helped increase (3–5 %) light interception in corn plants with a five to eightfold reduction in weed biomass (Begna et al. 2001). Planting wheat in a grid pattern with a row spacing of 2.5 cm improved grain yield to 9 % and reduced weed biomass to 30 % as compared to a row spacing of 12.8 cm (Weiner et al. 2001). In a similar study conducted to find out the effect of planting

Fig. 3.3 Late stage wheat in conventional management system

density (400 and 600 plants m^{-2}), row spacing (11, 18, and 23 cm) and mechanical weed control on wheat grain yield and protein content in the presence of condiment mustard (weed) showed opposite findings. There, the authors reported higher grain yield and protein contents in a wider-row system; however, they further reported that experiments conducted under reduced or weed-free conditions do not relate to wheat performance in organic management systems where weed severity is often a problem (Kolb et al. 2012). Sandhu et al. (2010) evaluated competitive ability of wheat against wild oat where row spacing of 15 cm along with high seed rate (150 kg ha^{-1}) significantly improved dry matter accumulation, LAI, PAR interception in wheat as compared to normal seeding rate (100 kg ha^{-1}) and row spacing (22.5 cm). Borger et al. (2010) manipulated row orientation [east-west (right angle to the sunlight) and north-south] in wheat, barely, canola, lupines, and field peas to suppress weeds and increase crop yield. The east-west orientation reduced weed biomass to 51 and 37 % and increased grain yield by 24 and 26 % in wheat and barley, respectively. They further elaborated that increases in grain yield in these crops and reduction in weed biomass is basically due to an increase in PAR interception by the crop canopy in east-west orientation. Yields of canola, field peas, and lupines were not altered by row orientation.

3.1.2.4 Cover Crops

Cover crops play an important role to suppress weeds and serve to maintain and improve soil productivity (reduction in soil erosion, improvement in soil structure

Fig. 3.4 Late stage wheat in organic management system

Fig. 3.5 A glimpse of wheat growing in an organic farm in Alberta, Canada

and organic matter). Possible reasons for weed suppression include shading and competition for available resources. Putnam et al. (1983) identified rye residues as one of the best mulches to control weeds for at least up to six weeks after rye desiccation. Similar results with respect to weed suppression were reported with the use of annual clover (*Trifolium incarnatum* L.) as compared to perennial clovers (*T. pretense* L. and *T. repens* L.) when used as spring cover crops (Nelson et al. 1991). However, management practices and genotypes played a major role in weed suppression in a study conducted with seven clover genotypes (Ross et al. 2001). In another study, wheat and rye debris significantly reduced weed emergence as compared to clover debris whose effect ranged from suppression to stimulation (Blum et al. 1997). Weed emergence and germination require specific amounts of light, temperature, and moisture; thus, cover crops can effectively compete with weeds for these environmental cues and act as a barrier to weed establishment (Sarrantonio and Gallandt 2003). Cover crops like *Hordeum vulgare, Secale cereale, Avena sativa, Brassica* spp. and *Melilotus officinalis* also may have crop-weed allelopathic interactions that inhibit germination, emergence, growth, and development of weed plants (Masiunas 1998). These allelopathic compounds either released by the plants or their decaying matter (crop residue) help to control weeds for a limited time (a few weeks) because they are rapidly degraded by the environmental conditions; however, their suppression effect may last for several months (Masiunas 1998). In this regard, soil microorganisms play an important role in converting inactive compounds from residues of cover crops to phytotoxic chemicals to control weeds. Cover cropping using legumes is one of best strategies to control weeds in organically managed lands because N recovery is at least 70–90 % higher with the use of leguminous crops than synthetic fertilizers (Smil 2000). Hairy vetch and red clover released half of their nitrogen within four weeks of their incorporation as green manure/residue and improved soil inorganic nitrogen similar to those when synthetic (NH_4NO_3) fertilizer was used at the rate of 179 kg ha^{-1} (Stute and Posner 1995). Moreover, soil nitrate concentration is positively correlated with weed germination and emergence (Karssen and Hilhorst 1992; Taylorson 1987); therefore, the use of mulch/residue/cover crops that release nitrogen slowly can help delay weed germination, emergence, establishment, and density.

3.1.2.5 Crop Rotation

Crop rotation is another technique to control weeds as some weed species are more associated with particular crop plants as compared to others. For example, wild oats are mostly seen in wheat and barley, and not in rice; whereas barnyard grass is mostly present in rice fields. The increase in population of such weeds will continue if the same crop is harvested year after year because environmental and cultural conditions favoring crop plants also favor such weed species. The population of such weeds can be more easily managed by growing various crops intermittently in sequence with other crops (crop rotation) that discourage the development of such weeds which are in close association with a particular crop. In Canada, it was documented that the population of *Setaria glauca* (yellow foxtail) was high in flax fields when it was followed by oat instead of maize and wheat (Kommedahl and Linck 1958). Similarly, crop rotations using alfalfa and

Fig. 3.6 Competition between wheat crop and different kinds of weeds in an organic field

Fig. 3.7 Example of a wheat variety with higher competitive ability

few other perennial forage crops were shown to decrease/control weeds to a larger extent in Canada (Entz et al. 2002). In another study, a sharp decline in the density of *Bromus tectorum* (downy brome) was detected using a winter wheat-rapeseed rotation rather than continuous wheat in Alberta, Canada from 1988–1993 (Blackshaw 1994).

3.1.2.6 Crop/Cultivar Mixture

Crop or cultivar mixtures can be used to utilize more resources as compared to weed plants. Weed populations can also be minimized using relay cropping. Crop mixtures/sequences/relay cropping should be designed to maximize resource utilization of crop plants and efficiently use of nutrients, water, and light from weed plants. In most cases, intercrops suppress weeds better than the sole crops (Liebman and Dyck 1993), increase yield stability and protect crop plants from various pests and diseases (Kaut et al. 2008; Pridham et al. 2007). In organic systems, crop mixtures of wheat and barley have higher or equal wheat grain yield as compared to wheat when sown in monoculture, and this yield increase was accredited to the weed suppressive ability of the barley cultivar used in the experiment (Kaut et al. 2008). Rodriguez (2006) used three different barley cultivars to study the effect of cultivar mixtures on weed suppression and reported that cultivar mixtures suppress weeds. However, suppressive ability varied among cultivars. Root development ability of barley cultivars was reported to be one of the main reasons of suppressiveness and on some occasions, cultivar blends did not consist of a cultivar with good inherent competitive ability. In a similar study, a cultivar mixture of two Canadian cultivars, AC Superb and AC Interpid in a ratio of 1:1 and 1:2 gave better results than AC Superb alone (Kaut et al. 2009). Nelson et al. (2012) studied monocultures and intercropping of wheat, barley, canola, and field pea on weed suppression ability. They reported that 50 % of intercrops suppressed weeds in an organic management system and barley in monoculture as well as in intercropping had the ability to suppress weed biomass.

3.1.3 Mechanical Weed Control

Physical removal of weeds either by preseeding tillage or hand-weeding/hoeing in crop is the oldest method of weed control. Pre-seeding tillage is often practiced on organic farms to uproot weedy plants, disturb weed seed emergence, enhance water infiltration, and improve soil structure. However, the disruption of soil structure and tilth by mechanical tillage may increase soil erosion. The effect of tillage practices to control weeds is highly contradictory. Few studies have presented high weed infestation in tilled plots (Roberts and Feast 1972) whereas others have documented more weeds in no-tilled plots (Moss 1985). Tillage practices vary considerably among crops (Buhler and Oplinger 1990), locations/soils (Buhler and Mester 1991), and environmental/weather conditions. The reduction in weed population by tillage practices also depends on the presence of weed seeds in the soil surface (shallow or deep) and the capacity of weed species to germinate from various soil depths. Teasdale et al. (2007) conducted a long-term (nine year) study to elucidate the benefits of no-tillage and organic cropping on weed control and soil improvement. The authors reported that reduced tillage under organically managed land may not provide satisfactory results with respect to weed control but it has long-term benefits to improve soil quality as compared to zero-tillage, which is often practiced in conventional management systems.

3.2 Genetics and Genomics of Competitiveness

Competitive ability is a genetic trait that varies among crop cultivars and within species as well as genus. Several studies have shown that barley is more competitive than rye, wheat, and flax (Bell and Nalewaja 1968; Fischer et al. 2001; Odonovan et al. 1985; Pavlychenko and Harrington 1934). Lemerle et al. (1995), however, ranked oat, cereal rye, triticale, oilseed rape, spring wheat, spring barley, and field pea in order of competitiveness against ryegrass (weed) in Australia. In the USA, rye was more competitive than barley and wheat (Nalewaja 1978) and rye was reportedly more competitive against wheat and other grain legumes in the UK (Millington et al. 1990). In another study (Seavers and Wright 1999), oat was found to be the most suppressive against *Galium aparine* followed by barley and wheat in the UK. Beres et al. (2010) reported that winter cereals exhibit better competitive ability against weeds than spring wheat in western Canada.

Various studies have been conducted to elucidate competitive ability within cultivars of different cereal crops like wheat (Huel and Hucl 1996), barley (Paynter and Hills 2009), rice (Zhao et al. 2006), and oat (Schaedler et al. 2009). Experiments on wheat competitive ability are well documented and have been conducted in different parts of the world including the UK (Cosser et al. 1997; Moss 1985; Richards 1983), India (Balyan et al. 1991; Mishra and Singh 2008), Pakistan (Ijaz and Hassan 2007), USA (Challaiah et al. 1986a; Wicks et al. 1986), Denmark (Christensen 1994), Germmany (Verschwele et al. 1993), Australia (Lemerle and Cooper 1996; Lemerle et al. 1996b, c), and Canada (Blackshaw et al. 2003, 2004; Huel and Hucl 1996; Kaut et al. 2008; Mason et al. 2007; Mason and Spaner 2006; Reid et al. 2009a, b; Reid et al. 2011). In winter wheat studies, Challaiah et al. (1986b) reported significant differences among cultivars and found "Turkey" to be a better competitor than a widely grown winter wheat cultivar "Centurk 78" against downy brome (*Bromus tectorum*). They reported a yield reduction of 9–21 % at one and 20–41 % at another location in various cultivars under study.

In Canada, various experiments/studies have been done on paired organic and conventional management systems to find out the differences in competitive ability of wheat genotypes against weeds. Kirkland and Hunter (1991) compared three cultivars of Canada Western Red Spring (CWRS) and Canada Prairie Spring Wheat (CPSW) classes, reporting Neepawa to be a better competitor and weed suppressor than HY320 and HY355. Huel and Hucl (1996) conducted experiments to study genotypic variation for competitive ability in spring wheat cultivars under simulated weed competition conditions for three years. Their results indicated that highest yielding genotypes are different under weedy and weed-free conditions. The cultivars that exhibited high competitive ability under weedy environments were comparatively taller with larger seedling ground cover and flag leaf length. They also found a yield reduction of 42 and 59 % and weed biomass reduction of 48 and 61 % at weed seeding rates of 48 and 96 seeds m^{-2}, respectively. One of the largest studies was undertaken by Reid et al. (2009b), who evaluated a random population of 79 F_6 derived recombinant inbred sister lines developed from a cross between AC Barrie, a CWRS wheat and the widely adapted CIMMYT cultivar Attila. The population was tested under organic and conventional management systems for three years and the results indicated differences in heritability esti- mates among systems (organic and conventional) for six out of 14 traits. Geno- types selected at 10 % selection intensity shared 50 % or fewer common lines in both management systems for nine traits including grain yield. They suggested that selection differences exist between the two systems and breeding spring wheat for organic farming should be conducted on organically managed land.

In summary, genetically based differences of competitive ability exist between crop species and also within cultivars, especially for wheat. This variation for competitive ability can be used to breed cultivars for improved competitiveness but greater genetic variation is required to improve this trait in wheat (Lemerle et al. 2001a, b). The heritability of competitive ability seems to be low or difficult to measure/record because competitiveness is a population phenomenon rather than a single plant. Therefore, low heritability makes it difficult to improve competitiveness through selective breeding; however, selection can be done on the basis of other traits like early season vigor and plant height to improve competitiveness.

In a study to evaluate weed suppression ability of 63 modern and historical wheat cultivars, Wissuwa et al. (2009) reported that top ranked cultivars sup- pressed weed weight (on a plot basis) by 573 % as compared to cultivars with lowest weed suppression ability. The study demonstrated the presence of large genetic variability among wheat cultivars for competitive ability against weeds. Mokhtari et al. (2002) investigated the genetic basis of variation for tolerance of two $F_{2:3}$ wheat populations for competition against *Lolium rigidum* (annual rye- grass). The population was derived from crosses using locally adapted Australian cultivars with poor and good competitive ability. The two populations (each consisted of 40 lines; first derived from two late flowering cultivars while second developed from two early flowering cultivars) were evaluated in the field with and without *L. rigidum*. The study reported significant levels of genetic variation in each population for tolerance against *L. rigidum* and reported heritability estimates of 0.25 and 0.57 for percentage of yield loss (tolerance) due to competition on an entry mean basis in early and late crosses, respectively. This suggests that that selection for competitive ability in F_3 or later generations can be effective. The study further revealed a strong relationship between 1) plant height and percentage yield loss and 2) leaf length and percentage yield loss; suggesting selection based on plant height and leaf length can be performed for low percent yield loss in wheat. In the population derived from early flowering parents, the effectiveness of indirect selection based on total dry weight, and number of heads per plants in monoculture was higher than direct selection. In late crosses, none of the trait had high heritability and correlation with tolerance that was required to achieve greater results from direct selection. Moreover, a positive correlation between time to anthesis and percentage yield losses in both populations indicated that earliness is linked to competitiveness/tolerance.

A multilocation (12 environments) testing of a spring wheat population on organic and conventional lands for three years revealed that selection differences were present between two management systems and across environments; thus selection/breeding for wheat grain yield should be conducted on organically managed lands (Reid et al. 2011). Coleman et al. (2001) also found variation for narrow sense heritability of agronomic traits (0.99 for flowering date, 0.90 for plant height at stem elongation, 0.34 for number of tillers, and 0.18 for crop dry matter during tillering) associated with weed suppression and concluded that estimates varied with respect to population and environmental conditions. Therefore, we can conclude that weed competitive ability is a complex trait with low heritability and indirect selection may be beneficial in some instances. The authors also found many quantitative trait loci (QTL) for various traits (anthesis, plant height, flag leaf size, and the size of the first two leaves) that mapped to similar positions for two years on chromosomes 2B and 2D. These QTL conferred competitively ability against weeds in 161 doubled haploid (DH) line's population derived from a cross between Cranbrook and Balbred. The time to anthesis QTL on chromosome 2D suggested it to be a photoperiod insensitive gene whereas the QTL position of plant height at anthesis on chromosome 4B could be an *Rht-B1b* (dwarfing) gene (Coleman et al. 2001).

Early season vigor in wheat is linked with greater seedling leaf area and longer coleoptiles (Spielmeyer et al. 2007). Early season vigor confers competitiveness in wheat plants against weeds (Bertholdsson 2005; Huel and Hucl 1996) and also helps to prevent water loss from the soil surface. Genetic variation for early season vigor is present in cultivated wheat (Richards and Lukacs 2002) and can be exploited to increase competitive ability against weeds. However, enhanced early season vigor via increased coleoptile length and early leaf area is limited in wheat populations carrying the *RhtB1b* dwarfing gene as it has a negative association with these traits (Rebetzke et al. 2001b). Spielmeyer et al. (2007) developed a recombinant inbred line (RIL) population, consisted of 460 lines, from a cross between Vigour 18 (exhibiting high seedling vigor) and Chuan-Mai18 (CM18) (carrying the GA-sensitive gene *Rht8*) to uncover genomic regions conferring greater seedling leaf area and longer coleoptiles in the absence of major dwarfing genes (*RhtB1b or RhtD1b*). The study reported a QTL on chromosome 6A that explained 6 and 14 % of the phenotypic variation for seedling leaf width and coleoptile length, respectively. The authors also reported the SSR marker "NW3106" associated with coleoptile length. Therefore, MAS for coleoptile length can be performed in early breeding generations using NW3106 to select lines for increased early season vigor. In a similar study, 85 different QTLs, on chromosomes 1D, 4D, and 7D, were reported for various traits (unit leaf area, leaf area ratio, biomass allocation, relative growth rate, and specific leaf area) on chromosomes 1, 4, and 7D; duration and rate of leaf elongation, cell length, and production rate on chromosome 2D; and growth rate of number of leaves and tillers along with total leaf mass on 5D chromosome, linked to early vigor in *Aegilops tauschii*, the D genome donor of hexaploid wheat (*Triticum aestivum*) (ter Steege et al. 2005). It was reported that most of the modern wheat cultivars contain dwarfing genes (Wojciechowski et al. 2009) such as *Rht-B1b* or *Rht-D1b* which reduced plant height and suppressed early season vigor and sensitivity to giberellic acid (Worthington and Reberg-Horton 2013). Therefore, breeding wheat cultivars for enhanced competitive ability can be performed with other/alternate dwarfing genes like *Rht-8* that reduce plant height without compromising early season vigor (Addisu et al. 2009; Rebetzke and Richards 1999). In this regards, *Ppd-D1a* was also reported to exhibit an even more strong positive influence on early season vigor in wheat (Addisu et al. 2009).

Enhanced nitrogen use efficiency (NUE) of wheat cultivars can also help to compete against weeds, especially in organic systems. Genetic variation for NUE exists in wheat suggesting the possibility of improving wheat NUE (May et al. 1991; Ortiz et al. 1997). An et al. (2006) used a DH population to find QTLs for traits related to nitrogen uptake in a field trial conducted with low and high nitrogen treatments. The authors reported eight and nine QTLs for nitrogen uptake under high and low nitrogen conditions, respectively, that can be deployed in breeding programs to increase early season vigor in wheat. Nitrogen use efficiency can also be improved by establishing an efficient plant growth promoting rhizosphere (PGPR) (Wissuwa et al. 2009) and arbuscular mycorrhizas (AMs) (Gosling et al. 2006). The PGRP improved NUE by protecting roots from attack by soil borne pathogens and maintaining effective mineralization-driven nutrient supply to plant root system (Cook 2007). The AM, a trait that has been reported as "rhizosphere competence", was termed essential for micronutrient, phosphorous, and efficient water uptake in crop plants grown under organic or low input conditions (Mäder et al. 2000). However, AM symbiosis is less effective to crop plants grown under optimal/high input conditions, especially when high levels of plant available phosphorous are present in the soil (Johnson et al. 1997). van Bueren et al. (2011) suggested that recent plant breeding programs aimed for high input farming system might have focused on selection against rhizosphere competence. Wheat cultivars developed before 1950 were more dependent on mycorrhizal symbiosis than later ones (Hetrick et al. 1993). Similarly, landraces of mycorrhizal wheat planted under low input (phosphorous) system yielded more grain yield than modern cultivars planted under similar conditions (Egle et al. 1999; Hetrick et al. 1992). It is therefore suggested that mycorrhizal responsive genes identified in wheat (Hetrick et al. 1995) need to be exploited to breed wheat cultivars for enhanced mycorrhizal root symbiosis to improve competitive ability against weeds.

Environmental conditions also affect competitive ability of crop plants and weeds. Various traits appear to confer different competitive behavior to control weeds in different environments. It was found that few cultivars exhibit environment specific competitive advantages (Lemerle et al. 2001b). Cultivars with better stability in response to changing environmental conditions, especially in terms of their suppression behavior against weeds, will be more desirable to the breeders breeding for organic lands. It is suggested that studies on large populations with QTL analysis for each year in paired/separate management (organic *vs* conventional) system will broaden breeding activities and help to identify molecular markers closely linked with gens/QTLS conferring competitive ability. These molecular markers can be deployed in plant breeding programs through marker assisted selection (MAS) to increase selection and breeding efficiency. Studies conducted to find QTL for various traits conferring competitive ability are presented in Table 3.1.

Weed species also vary in competitive ability and it is a well-established fact that grassy weeds are more aggressive than others. It was reported that 120 plants m^{-2} of *Avena sterilis* L. had greater interference on barley than *Phalaris minor* at 400 plants m^{-2} (Dhima et al. 2000). Weed species were also ranked as *Chenopodium album* L. > *Phalaris minor.* > *Sinapis arvensis* L. according to their order of competitive ability (Iqbal and Wright 1999). However, similar competitive ability of *P. minor* and *P. brachystachys* has been reported in wheat, with the former being more aggressive with faster growth rate (Afentouli and Eleftherohorinos 1996). *Sinapis alba* (white mustard) was reported to be a stronger competitor than *L. multiflorum* and *C. album* (Olsen et al. 2006). The competitive behavior of weeds is highly dependent on the growth rate of their roots (Wicks et al. 1986).

The knowledge of traits conferring competitiveness is very important for cereal breeders in making selections and developing new cultivars for weed suppression. Although plant breeders select for a number of quality, disease, and agronomic traits, priority has never been given to competitive ability. Selections are often performed in conventional systems where weeds are controlled with herbicides. Selections made in high input systems may not suit well to organically managed lands because adaptability to abiotic (low nutrients) and biotic (diseases and weeds) stresses is more desirable on organic lands. In organic systems, genotype by environment interaction is greater due to existence of weeds in a larger number and low nutrients that make selection challenging for traits conferring competitiveness (Coleman et al. 2001). Additionally, this selection can be at the expense of other important traits required to develop a new cultivar (Brennan et al. 2001). The existence of genetic variability is a prerequisite for a trait that needs an improvement. Therefore, an understanding of the inheritance, genetic variability, and genetics of traits conferring competitive ability can help to design an efficient breeding scheme and potentially a marker assisted selection (MAS) strategy to breed cultivars for organically managed systems.

Recent advancements in the field of biotechnology especially relating to DNA have broadened the prospects of selecting suitable/desirable genotypes based on DNA markers. The conventional plant breeding, significantly influenced by the environmental conditions, involves the selection of superior genotypes based on the phenotypes followed by their commercial cultivation. In order to meet the food

Table 3.1 Quantitative trait loci (QTL) conferring competitive ability

Crop	Parents/Population type	Trait	Chromosome	QTL	Linked markers	Variation explained (%)	Reference
Wheat	Hanxuan 10 × Lumai14 (DH)	Nitrogen uptake	2D		WMC179.1–WMC256	21.9	(An et al. 2006)
			6B		P3454–165–P3516–205	10.9	
			5B		WMC363–WMC376	12.4	
			7D		Xgdm88–WMC463	10.1	
			5A		Xgwm595–WMC410	15.9	
			2D		Xgwm157–Xgwm539	14	
			4B		Xgwm495–Xgwm149	14.1	
Wheat	Chuan-Mai18 × Vigour18 (RIL)	Coleoptile length	6A		NW3106	8	(Spielmeyer et al. 2007)
		Seedling leaf width				14	
		Plant height				16	
Wheat	No. 22 Xiaoyan × 92517-25-1	Weed competitive ability	1A	QWca1A-2	Xbarc164–Xbarc108	6.46	(Zuo et al. 2012)
			1A	QWca1A-2	Xbarc49–Xbarc253	8.98	
			2B	QWca2B-1	Xgdm113–Xgwm194	10.58	
			5D	QWca5D-1	Xgdm115–Xgwm639	16.24	
Wheat	No. 22 Xiaoyan × 92517-25-1	Allelopathic potential	2B	QAp2B-3	Xgdm113–Xgwm194	7.24	(Zuo et al. 2012)
			2B	QAp2B-3	Xbarc362–Xbarc350	11.76	
			1A	QWca1A-2	Xbarc164–Xbarc108	6.64	
Wheat	Berkut × Krichauf	Plant height	6B	Qf.6B	gwm219/cfd190	14.7A	(Trethowan et al. 2012)
			2A	Qh.2A1	gwm71/wPt5647	16.4A	
			2A	Qh.2A2	wPt3114/gwm95	22.2A	
			2A	Qh.2A1	gwm71/wPt5647	16.8A	
			2A	Qh.2A1	gwm71/wPt5647	15.1A	
			2D	Qh.2D	cfd44/wPt9797	13.4A	
			4A	Qh.4A	wmc106/wmc-48b	17.0A	
			6A	Qh.6A	wPt7063/wmc807	25.5A	
			6A	Qh.6A	wPt7063/wmc807	24.7A	
			6A	Qh.6A	wPt7063/wmc807	24.7A	

(continued)

Table 3.1 (continued)

Crop	Parents/Population type	Trait	Chromosome	QTL	Linked markers	Variation explained (%)	Reference
Wheat	Huapei 3 × Yumai 57	Lodging	1B	qLdg1B	Xcwem6.1–Xwmc128	−0.33	(Zhang et al. 2008)
			2B	qLdg2B	Xbarc129.1–Xgwm111	−0.25	
			3A	qLdg3Ad	Xbarc51–Xbarc157.1	0.33	
			4B	qLdg4B	Xwmc48–Xbarc1096	−0.24	
			4D	qLdg4Db	Xbarc334–Xwmc331	0.28	
Wheat	Cranbrook × Halberd	Plant height	4BS		XcsME1	26	(Rebetzke et al. 2001a)
			4BL		XksuC2	22	
			5AL		P35/M54-6	11	
			2DS		Xwmc018	11	
			4BS		XcsME1	37 (46)	
			4BL		XksuC2	22 (28)	
			4BS		XcsME1	42 (49)	
			4BL		XksuC2	19 (22)	
			3AL		Xwmc050	10 (12)	
			4BS		XcsME1	24 (30)	
			4BL		XksuC2	11 (14)	
			5AL		P32/M51-7	10 (13)	

(continued)

Table 3.1 (continued)

Crop	Parents/Population type	Trait	Chromosome QTL	Linked markers	Variation explained (%)	Reference
Wheat	Cranbrook × Halberd	Coleoptile length	4BS	XcsME1	38 (45)	(Rebetzke et al. 2001a)
			4BL	XksuC2	23 (27)	
			5AL	P31/M58-2	11 (13)	
Wheat	Cranbrook × Halberd	Seedling internode length	4BS	XcsME1	61 (68)	(Rebetzke et al. 2001a)
			4BL	XksuC2	34 (38)	
			3BL	XksuG59	11 (12)	
Wheat	Cranbrook × Halberd	Leaf 1 length	4BS	XcsME1	49 (54)	
			4BL	XksuC2	31 (34)	
Wheat	Cranbrook × Halberd	Leaf 1 length	2DL	Xwmc112	15 (28)	
			5AL	P38/M54-8	12 (23)	
		Leaf 1 width	5BL	P38/M51-1	13 (42)	
			5BL	P35/M55-2	11 (35)	
Wheat	Cranbrook × Halberd	Coleoptile tiller size	4BL	Xgwm495	22 (39)	
			5AL	P31/M58-2	10 (18)	
Wheat	Cranbrook × Halberd	Anthesis (1998)	2D		16.8	(Coleman et al. 2001)
			2B		14.2	
		Anthesis (1999)	2D		20.2	
			2B		12.1	
		Crop Dry matter 1	4B		11.5	
		Crop Dry matter 2	1A		9.5	
Wheat	Cranbrook × Halberd	Flag leaf Area	2B		13.4	
			2D		11.3	
			2A		8.7	
		Flag leaf Length	2B		11.1	
			2D		10.6	
		Flag leaf Width	2D		12.3	
			2D		10.4	

(continued)

Table 3.1 (continued)

Crop	Parents/Population type	Trait	Chromosome	QTL	Linked markers	Variation explained (%)	Reference
Wheat	Cranbrook × Halberd	Canopy height	2D			13.8	
			2B			11.8	
			2D			17	
			2B			12.4	
Wheat	Cranbrook × Halberd	Plant height	4B			29.5	
			3A			13.1	
			5A			12	
			4B			27	
			3A			13.2	
			5A			10	
Wheat	Cranbrook × Halberd	Leaf 1 Area	2D			11.2	
		Leaf 1 Length	2D			10.6	
		Leaf 2 Length	2D			10.2	
			2D			11.5	
		Leaf 2 Width	2D			9	
		Leaf Area Index	7B			9.4	
			7B			11.7	
Wheat	Cranbrook × Halberd	Tiller number	2B			17.9	
			4A			14.6	
		Yield	3A			12.2	
			3B			9.8	
		Thousand Grain Weight	5A			11	
			2D			8.4	
			5A			12	
			2B			9.9	
		Head density	2D			13.6	

(continued)

Table 3.1 (continued)

Crop	Parents/Population type	Trait	Chromosome	QTL	Linked markers	Variation explained (%)	Reference
Wheat	Sunco × Tasman	Allelopathy	2B		P32/M48-316	29	(Wu et al. 2003)
			2B		P32/M48-93		
Wheat	Cranbrook × Halberd	Coleoptile length	1AS		ksuG9c	3.6	(Rebetzke et al. 2007)
			2BS		wPt-0615	2.7	
			2DS		Stm55ltgag	5.9	
			3BS		wPt-8855	2.8	
			4BS		Rht-B1b	35.4	
			5AL		psr426	8.6	
			5DS		psr326b	4.1	
			Ep1:		Rht-B1b,wmc76	2.1	
			4BS*7BS				
			Ep2:		Rht-B1b,wPt-6463	1.6	
			4BS*7BS				
			Ep3:		stm55ltgag,RGA74.15	2.7	
			2DL*5BS				
Wheat	Sunco × Tasman	Coleoptile length	2BS		wmc474	2.8	(Rebetzke et al. 2007
			2DS		gwm515c	1.2	
			4AS		gwm165	2.3	
			4BS		Rht-B1b	20.6	
			4DS		Rht-D1b	29.0	
			6BL		barc178	2.2	
Wheat	CD87 × Katepwa	Coleoptile length	2DS		P36/M43-1	3.8	(Rebetzke et al. 2007
			3BS		abg75c	3.6	
			4BS		Rht-B1b	28.2	
			5DS		abg3a	2.8	
			6BL		P35/M39-9	4.1	
			7AL		KsuH9c	4.4	
			Ep1:		bcd310,P41/M51-2	0.5	
			1AL*5DS				

(continued)

Table 3.1 (continued)

Crop	Parents/Population type	Trait	Chromosome	QTL	Linked markers	Variation explained (%)	Reference
Wheat	Kukri × Janz	Coleoptile length	4AS		*gwm637*	0.3	(Rebetzke et al. 2007)
			4BS		*Rht-B1b*	22.5	
			4DS		*Rht-D1b*	27.3	
			6BL		*gwm219*	6.7	
Wheat	Cv. No. 22 Xiaoyan × Cv. 92517-25-1	Allelopathic potential	1A	QAp1A-1	Xbarc164–Xbarc108	12.56	(Zuo et al. 2012)
			2B	QAp2B-3	Xbarc334–Xgwm129	10.69	
			2B	QAp2B-3	Xgdm113–Xgwm194	7.24	
			2B	QAp2B-3	Xbarc362–Xbarc350	11.76	
Wheat	Cv. No. 22 Xiaoyan × Cv. 92517-25-1	Weed competitive ability	1A	QWca1A-2	Xbarc164–Xbarc108	6.46	
			1A	QWca1A-2	Xbarc49–Xbarc253	8.98	
			2B	QWca2B-1	Xgdm113–Xgwm194	10.58	
			5D	QWca5D-1	Xgdm115–Xgwm639	16.24	
Wheat	Cv. No. 22 Xiaoyan × Cv. 92517-25-1	Thousand grain weight	5D	QTgw5D-2	Xbarc235–Xbarc299	5.75	
			5D	QTgw5D-2	Xgwm639–Xwmc776		
Rice	Zhenshan 97 (Indica) × IRAT109 (upland)	Flag leaf length	3	QFll3	RM523–RM231	7.69	(YUE et al. 2006)
			2	QFll2	RM262–MRG0303	19.69	
			4	*QFll4*	RM255–RM349	7.75	
			4	*QFll4*	*RM255–RM349*	11.98	
			5	QFll5	RM480–RM334	5.32	
			7	QFll7	RM274–RM480	8.41	
			10	QFll10	RM596–RM271	5.95	
Rice	Zhenshan 97 (Indica) × IRAT109 (upland)	Flag leaf area	2	QFla2	*RM262–MRG0303*	10.47	(YUE et al. 2006)
			3	QFla3	RM523–RM231	4.41	
			4	*Qfla4*	RM255–RM349	26.83	
			4	*Qfla4*	*RM255–RM349*	17.68	
			5	*Qfla5*	RM274–RM480	9.22	
			6	*Qfla6*	*RM111RM276*	11.8	
			10	*Qfla10*	RM596-RM271	5.68	

(continued)

Table 3.1 (continued)

Crop	Parents/Population type	Trait	Chromosome	QTL	Linked markers	Variation explained (%)	Reference
Rice	Zhenshan 97 (Indica) × IRAT109 (upland)	Flag leaf width	4	*Qflw4*	RM255–RM349	35.81	(YUE et al. 2006)
			4	*Qflw4*	*RM255–RM349*	16.5	
			5	*Qflw5*	RM421–RM274	13.4	
			5	*Qflw5*	RM274–RM480	10.08	
			6	*Qflw6*	RM539–RM527	4.41	
Rice	Lemont × Teqing	Coleoptile emergence	1	qFV-1-1	RG140–RM259	3.50A	(Zhou et al. 2007)
		Seedling emergence	1	qFV-1-1	RG140–RM259	3.75A	
		Seedling height	1	qFV-1-2	CDO118–RG462	3.59A	
		Seedling height	2	qFV-2 2	RG555–OSR17	0.325	
		Seedling height	3	qFV-3-1	RG341b–RM218	−0.326	
		Coleoptile emergence	3	qFV-3-2	OSR31–RM168	4.73	
		Seedling emergence	3	qFV-3-2	OSR31–RM168	5.09	
		Seedling emergence	3	qFV-3-3	RM148–RM85	−3.4	
		Seedling emergence	5	qFV-5-1	RM13–N10150–Y1049	−3.74	

(continued)

Table 3.1 (continued)

Crop	Parents/Population type	Trait	Chromosome	QTL	Linked markers	Variation explained (%)	Reference
Rice	Lemont × Teqing	Seedling height	5	qFV-5-1	RM13–N10150–Y1049	-0.24	(Zhou et al. 2007)
		Dry weight	5	qFV-5-1	RM13–N10150–Y1049	-3.75	
		Seedling height	5	qFV-5-2	RM161–CDSR49	-0.51	
		Seedling height	10	qFV-10	C223–RM228	-0.42	
		Dry weight	10	qFV-10	C223–RM228	-3.12	
Rice	Lemont × Teqing	Shoot length	2		RM250–RG256	-1.55	(Zhang et al. 2005)
			3		RM148–RM85	-1.83	
			7		G20–RG678	1.91	
			8		RM230–RM264	-1.31	
		Dry matter weight	5		N10150–Y1049	-0.16	
			7		RG678–C285	0.08	
Rice	Lemont × Teqing	Plant height	1	qPh1	RM104–RZ14	2.25	(Zhu et al. 2008)
			2	qPh2	OSR26–RM208	-1.83	
			3	qPh3	RG348–C636x	-1.96	
			6	qPh6	RM30–RM340	-2.41	
			8	qPh8a	RM223–RM210	2.32	
			8	qPh8b	OSR7–RM230	-2.98	
			10	qPh10	RZ400–RM258	3.34	
			12	qPh12	RG20q–RG901	-2.37	
Rice	Vandana/Moroberekan	Plant height	1	qPH-1	RG811–BP127	2.98	(Zhang et al. 2004)
			2	qPH-2	RG324–RG83	8.38	
			3	qPH-3-1	RG409a–RG266	5.06	
			3	qPH-3-2	RG369–RG745	-3.94	
			5	qPH-5	G387A–G360b	-5.59	

demand of ever increasing human population, plant breeders are adopting new approaches to develop cultivars with improved crop yields, better quality, and enhanced resistance to biotic and abiotic stresses.

During the last two decades, a large number of genetic tools based on molecular marker technologies have been evolved that resulted in precise genotyping to study the level and change of molecular diversity in a particular crop species. Various types of DNA markers have developed including (a) Polymerase chain reaction (PCR) based such as amplified fragment length polymorphisms (AFLPs), random amplified polymorphic DNAs (RAPDs), and microsatellites or simple sequence repeats (SSRs); (b) hybridization based such as restriction fragment length polymorphisms (RFLPs); (c) single nucleotide polymorphisms (SNPs); and (d) plant retrotransposons. The characteristics of these markers and their application to wheat breeding have been extensively reviewed (Alexandrova et al. 1999; Ganal and Roeder 2007; Gupta et al. 1999; Korzun and Ebmeyer 2003; Langridge et al. 2001; Röeder et al. 2005). Marker Assisted Selection (MAS) based on DNA markers has greatly improved the efficiency of conventional plant breeding programs throughout the world by detecting and characterizing genetic variation among the available germ plasm (Tanksley and McCouch 1997) and identifying/pyramiding genes/Quantitative Trait Loci (QTL) of interest using genetic transformation or marker assisted techniques (Salvi and Tuberosa 2005; Tanksley and McCouch 1997). Marker-assisted selection can be defined as an indirect selection process where a trait of interest is selected not on the basis of its phenotype but with the help of a marker (morphological, biochemical, DNA based) linked to the trait of interest. Marker assisted selection based on molecular markers is being used to increase selection efficiency in plant breeding programs because molecular markers have several advantages over traditional methods such as (i) molecular markers are not influenced by environment and growth stage of the plan,t (ii) molecular markers can facilitate the quick transfer of recessive genes into desirable genotype, (iii) possibility of distinguishing between homozygotes and heterozygotes (in case of codominant markers), (iv) identifying uniformity and distinctness in case of cultivar development by providing unique DNA profiles and (v) molecular markers provides opportunity to select genotypes for traits which are difficult or impossible to select, e.g., when natural pathogen infestation is not available and artificial infestation is prohibited in crop breeding programs aimed for resistant breeding.

Although molecular markers are simply diagnostic tools to increase selection efficiency and are not directly interfering or altering the genome at DNA level but their use in breeding for organic agriculture has been recently debated in a European plant breeding workshops on the use of molecular markers (van Bueren et al. 2005). The participants were of the view that MAS can only be a useful breeding tool when trait of interest is monogenic and for complex traits related to organic breeding such as nutrient use efficiency (NUE) and competitive ability, molecular markers would not be available in the near future. Phenotypic selection was considered to be one of the best way to make selections for organic breeding (van Bueren et al. 2005). However, the use of MAS as a non-GMO (genetically modified organisms) strategy to breed cultivars for organic agriculture cannot be ignored (van Bueren et al. 2010). The authors also reported that molecular markers developed under and for conventional management systems may not be effective, when applied in organic context. In this regards, a study was conducted to map QTL affecting various agronomic and quality traits and to investigate the feasibility of organic wheat breeding by determining genetic differentials along with the effect of *Rht-B1* in paired organic and conventional management systems and the results indicated that heritability, correlation coefficients, effect of *Rht-B1* and selection differences exist between two systems. The study identified and mapped 45 QTL for various traits across organic and conventional management systems for three year and most of the identified QTL were specific to the management system. Moreover, consistent QTL mapped in both systems for grain yield, test weight, kernel weight, and days to flowering also differ with respect to their phenotypic variation and additive effects leading to the conclusion that QTL express differently in different environments and this is also true for organically managed systems (M. Asif, unpublished).

Based on the above discussion, it can be concluded that MAS can definitely improve breeding efficiency because it enables breeders to work with smaller populations in the field conditions and can efficiently improve traits which are controlled by single genes. Therefore, molecular markers are tools to complement phenotypic selections and there is dire need to balance between molecular and nonmolecular methods by paying attention to the opportunities to design methods to improve phenotypic selection.

References

Addisu M, Snape JW, Simmonds JR, Gooding MJ (2009) Reduced height (*Rht*) and photoperiod insensitivity (*Ppd*) allele associations with establishment and early growth of wheat in contrasting production systems. Euphytica 166:249–267

Afentouli CG, Eleftherohorinos IG (1996) Littleseed canarygrass (*Phalaris minor*) and short-spiked canarygrass (*Phalaris brachystachys*) interference in wheat and barley. Weed Sci 44:560–565

Alexandrova NA, Todorovska EG, Marinova EI, Atanassov AI (1999) DNA markers and their application in plant breeding for disease resistance in wheat. Bulg J Agri Sci 5:551–560

Alkamper J (1976) Influence of weed infestation on effect of fertilizer dressings. Pflanzenschutz-Nachrichten Bayer 29:191–235

An DG, Su JY, Liu QY, Zhu YG, Tong YP, Li JM, Jing RL, Li B, Li ZS (2006) Mapping QTLs for nitrogen uptake in relation to the early growth of wheat (*Triticum aestivum* L.). Plant Soil 284:73–84

Bailey KL (2010) Canadian innovations in microbial biopesticides. Can J Plant Pathol 32:113–121

Balyan RS, Malik RK, Panwar RS, Singh S (1991) Competitive ability of winter-wheat cultivars with wild oat (*Aavena udoviciana*). Weed Sci 39:154–158

Barberi P (2002) Weed management in organic agriculture: are we addressing the right issues? Weed Res 42:177–193

Begna SH, Hamilton RI, Dwyer LM, Stewart DW, Cloutier D, Assemat L, Foroutan-Pour K, Smith DL (2001) Weed biomass production response to plant spacing and corn (*Zea mays*) hybrids differing in canopy architecture. Weed Technol 15:647–653

Bell AR, Nalewaja JD (1968) Competition of wild oat in wheat and barley. Weed Sci 16:505–508

Beres BL, Harker KN, Clayton GW, Bremer E, Blackshaw RE, Graf RJ (2010) Weed-competitive ability of spring and winter cereals in the northern great plains. Weed Technol 24:108–116

Bertholdsson NO (2005) Early vigour and allelopathy: two useful traits for enhanced barley and wheat competitiveness against weeds. Weed Res 45:94–102

Blackshaw RE (1994) Rotation affects downy brome (*Bromus tectorum*) in winter-wheat (*Triticum aestivum*). Weed Technol 8:728–732

Blackshaw RE, Brandt RN, Janzen HH, Entz T (2004) Weed species response to phosphorus fertilization. Weed Sci 52:406–412

Blackshaw RE, Brandt RN, Janzen HH, Entz T, Grant CA, Derksen DA (2003) Differential response of weed species to added nitrogen. Weed Sci 51:532–539

Blum U, King LD, Gerig TM, Lehman ME, Worsham AD (1997) Effects of clover and small grain cover crops and tillage techniques on seedling emergence of some dicotyledonous weed species. Am J Alter Agri 12:146–161

Borger CPD, Hashem A, Pathan S (2010) Manipulating crop row orientation to suppress weeds and increase crop yield. Weed Sci 58:174–178

Brennan JP, Lemerle DPM (2001) Economics of increasing wheat competitiveness as a weed control weapon. Contributed paper presented to the 45th annual conference of the Australian agricultural and resource economics society, Adelaide, 2001

Buchanan GA, Hauser EW (1980) Influence of row spacing on competitiveness and yield of peanuts (*Arachis hypogaea*). Weed Sci 28:401–409

Buhler DD, Mester TC (1991) Effect of tillage systems on the emergence depth of giant (*Setaria faberi*) and green foxtail (*Setaria viridis*). Weed Sci 39:200–203

Buhler DD, Oplinger ES (1990) Influence of tillage systems on annual weed density and control in solid-seeded soybean (*Glycine max*). Weed Sci 38:158–165

Carlson HL, Hill JE (1985) Wild oat (*Avena fatua*) competition with spring wheat: plant density effects. Weed Sci 33:176–181

Challaiah Burnside OC, Wicks GA, Johnson VA (1986a) Competition between winter-wheat (*Triticum aestivum*) cultivars and downy brome (*Bromus tectorum*). Weed Sci 34:689–693

Challaiah O, Ramsel RE, Wicks GA, Burnside OC, Johnson VA (1986b) Evaluation of the weed competitive ability of winter wheat cultivars. In: Proceedings of the North central weed control conference tasmanian weeds society, Hobart, pp 85–91

Christensen S (1994) Crop-weed competition and herbicide performance in cereal species and varieties. Weed Res 34:29–36

Coleman RD, Gill GS, Rebetzke GJ (2001) Identification of quantitative trait loci for traits conferring weed competitiveness in wheat (*Triticum aestivum* L.). Aust J Agric Res 52:1235–1246

Cook RJ (2007) Management of resident plant growth-promoting rhizobacteria with the cropping system: a review of experience in the US Pacific Northwest. Eur J Plant Pathol 119:255–264

Cosser ND, Gooding MJ, Thompson AJ, FroudWilliams RJ (1997) Competitive ability and tolerance of organically grown wheat cultivars to natural weed infestations. Ann Appl Biol 130:523–535

Dhima KV, Eleftherohorinos IG, Vasilakoglou IB (2000) Interference between *Avena sterilis*, *Phalaris minor* and five barley cultivars. Weed Res 40:549–559

Ditomaso JM (1995) Approaches for improving crop competitiveness through the manipulation of fertilization strategies. Weed Sci 43:491–497

Doll H, Holm U, Sogaard B (1995) Effect of crop density on competition by wheat and barley with *Agrostemma githago* and other weeds. Weed Res 35:391–396

Downing RG (1921) Thick or thin seeding for wheat? Agricultural Gazette of New South Wales, p 205

Egle K, Manske G, Romer W, Vlek PLG (1999) Improved phosphorus efficiency of three new wheat genotypes from CIMMYT in comparison with an older Mexican variety. J Plant Nutr Soil Sci-Z Pflanzenernahrung Bodenkunde 162:353–358

Entz MH, Baron VS, Carr PM, Meyer DW, Smith SR, McCaughey WP (2002) Potential of forages to diversify cropping systems in the northern Great Plains. Agron J 94:240–250

Fischer AJ, Ramirez HV, Gibson KD, Pinheiro BD (2001) Competitiveness of semidwarf upland rice cultivars against palisadegrass (*Brachiaria brizantha*) and signalgrass (*B. decumbens*). Agron J 93:967–973

Ganal M, Röder M (2007) Microsatellite and SNP markers in wheat breeding. In: Varshney R, Tuberosa R (eds) Genomics-assisted crop improvement. Genomics applications in crops, vol 2. Springer, pp 1–24

Gosling P, Hodge A, Goodlass G, Bending GD (2006) Arbuscular mycorrhizal fungi and organic farming. Agric Ecosyst Environ 113:17–35

Gupta PK, Varshney RK, Sharma PC, Ramesh B (1999) Molecular markers and their applications in wheat breeding. Plant Breed 118:369–390

Harris P, Peschken D, Milroy J (1969) The status of biological control of the weed *Hypericum perforatum* in British Columbia. Can Entomol 101:1–15

Hashem A, Radosevich SR, Roush ML (1998) Effect of proximity factors on competition between winter wheat (*Triticum aestivum*) and Italian ryegrass (*Lolium multiflorum*). Weed Sci 46:181–190

Hetrick BAD, Wilson GWT, Cox TS (1992) Mycorrhizal dependence of modern wheat-varieties, landraces, and ancestors. Can J Bot 70:2032–2040

Hetrick BAD, Wilson GWT, Cox TS (1993) Mycorrhizal dependence of modern wheat cultivars and ancestors—a synthesis. Can J Bot 71:512–518

Hetrick BAD, Wilson GWT, Gill BS, Cox TS (1995) Chromosome location of mycorrhizal responsive genes in wheat. Can J Bot 73:891–897

Huel DG, Hucl P (1996) Genotypic variation for competitive ability in spring wheat. Plant Breed 115:325–329

Ijaz AK, Hassan G (2007) Competitive ability of various wheat cultivars with wild oats. African crop science conference proceedings. African crop science society, El-Minia, pp 1901–1904

Imaizumi S, Nishino T, Miyabe K, Fujimori T, Yamada M (1997) Biological control of annual bluegrass (*Poa annua* L.) with a Japanese isolate of *Xanthomonas campestris* pv. *poae* (JT-P482). Biol Contr Theory Appl Pest Manag 8:7–14

Iqbal J, Wright D (1999) Effects of weed competition on flag leaf photosynthesis and grain yield of spring wheat. J Agri Sci 132:23–30

Johnson NC, Graham JH, Smith FA (1997) Functioning of mycorrhizal associations along the mutualism-parasitism continuum. New Phytol 135:575–586

Karssen CM, Hilhorst HWM (1992) Effect of chemical environment on seed germination. In: Fenner M (ed) Seeds: the ecology of regeneration in plant communities. CAB International, Wallingford, pp 327–348

Kaut A, Mason HE, Navabi A, O'Donovan JT, Spaner D (2008) Organic and conventional management of mixtures of wheat and spring cereals. Agron Sustain Dev 28:363–371

Kaut AHEE, Mason HE, Navabi A, O'Donovan JT, Spaner D (2009) Performance and stability of performance of spring wheat variety mixtures in organic and conventional management systems in western Canada. J Agric Sci 147:141–153

Khan M, Donald WW, Prato T (1996) Spring wheat (*Triticum aestivum*) management can substitute for diclofop for foxtail (*Setaria* spp) control. Weed Sci 44:362–372

Kirkland KJ (1993) Weed management in spring barley (*Hordeum vulgare*) in the absence of herbicides. J Sustain Agri 3:95–104

Kirkland KJ, Hunter JH (1991) Competitiveness of canada prairie spring wheats with wild oat (*Avena fatua L*). Can J Plant Sci 71:1089–1092

Kolb LN, Gallandt ER, Mallory EB (2012) Impact of spring wheat planting density, row spacing, and mechanical weed control on yield, grain protein, and economic return in Maine. Weed Sci 60:244–253

Kommedahl T, Linck AJ (1958) The ecological effects of different preceding crop plants on *Setaria glauca* in flax. Proc Minnesota Acad Sci 25(26):91–94

Korres NE, Froud-Williams RJ (1997) The use of varietal selection and seed rates for enhanced weed suppression in winter wheat (*Triticum aestivum* L). In: Proceedings of the Brighton crop protection conference - weeds, pp 667–668

Korzun V, Ebmeyer E (2003) Molecular markers and their applications in wheat breeding. In: Pogna NE, Romano M, Pogna EA, Galterio G (eds) Proceedings of the 10th international wheat genetics symposium, Istituto Sperimentrale per la Cerealicoltura, Rome, pp 140–143

Langridge P, Lagudah ES, Holton TA, Appels R, Sharp PJ, Chalmers KJ (2001) Trends in genetic and genome analyses in wheat: a review. Aust J Agric Res 52:1043–1077

Lemerle D, Verbeek B, Coombes N (1995) Losses in grain yield of winter crops from *Lolium rigidum* competition depend on crop species, cultivar and season. Weed Res 35:503–509

Lemerle D, Cooper K (1996) Comparative weed suppression by triticale, cereal rye and wheat. In: Guedes-Pinto H, Darvey N, Carnide V (eds) Triticale: Today and Tomorrow. Springer, Netherlands, pp 749–750

Lemerle D, Verbeek B, Cousens RD, Coombes NE (1996a) The potential for selecting wheat varieties strongly competitive against weeds. Weed Res 36:505–513

Lemerle D, Verbeek B, Martin P (1996b) Breeding wheat cultivars more competitive against weeds. In: Brown H, Cussans GW, Devine MD, Duke SO, FernandezQuintanilla C, Helweg A, Labrada RE, Landes M, Kudsk P, Streibig JC (eds) Proceedings of the second international weed control congress, vols 1–4, pp 1323–1324

Lemerle D, Yuan TH, Murray GM, Morris S (1996c) Survey of weeds and diseases in cereal crops in the southern wheat belt of New South Wales. Aust J Exp Agric 36:545–554

Lemerle D, Gill GS, Murphy CE, Walker SR, Cousens RD, Mokhtari S, Peltzer SJ, Coleman R, Luckett DJ (2001a) Genetic improvement and agronomy for enhanced wheat competitiveness with weeds. Aust J Agric Res 52:527–548

Lemerle D, Verbeek B, Orchard B (2001b) Ranking the ability of wheat varieties to compete with *Lolium rigidum*. Weed Res 41:197–209

Liebman M, Robichaux RH (1990) Competition by barley and pea against mustard: effects on resource acquisition, photosynthesis and yield. Agric Ecosyst Environ 31:155–172

Liebman M, Dyck E (1993) Crop-rotation and intercropping strategies for weed management. Ecol Appl 3:92–122

Liebman M, Davis AS (2000) Integration of soil, crop and weed management in low-external-input farming systems. Weed Res 40:27–47

Mäder P, Edenhofer S, Boller T, Wiemken A, Niggli U (2000) Arbuscular mycorrhizae in a long-term field trial comparing low-input (organic, biological) and high-input (conventional) farming systems in a crop rotation. Biol Fertil Soils 31:150–156

Magdoff F (1995) Soil quality and management. In: Altiieri MA (ed) The science of sustainable agriculture. Westview Press, Boulder, pp 349–364

Masiunas JB (1998) Production of vegetables using cover crop and living mulches: a review. J Veg Crop Prod 4:11–31

Mason HE, Navabi A, Frick BL, O'Donovan JT, Spaner DM (2007) The weed-competitive ability of Canada western red spring wheat cultivars grown under organic management. Crop Sci 47:1167–1176

Mason HE, Spaner D (2006) Competitive ability of wheat in conventional and organic management systems: a review of the literature. Can J Plant Sci 86:333–343

May L, Vansanford DA, Mackown CT, Cornelius PL (1991) Genetic-variation for nitrogen use in soft red × hard red winter-wheat populations. Crop Sci 31:626–630

Millington S, Stopes C, Woodward L (1990) Rotational design and the limits of organic systems– the stockless organic farm? Monograph—British crop protection council, 163–173

Mishra JS, Singh VP (2008) Integrated weed management in dry-seeded irrigated rice (*Oryza sativa*). Indian J Agron 53:299–305

Mokhtari S, Galwey NW, Cousens RD, Thurling N (2002) The genetic basis of variation among wheat F-3 lines in tolerance to competition by ryegrass (*Lolium rigidum*). Euphytica 124:355–364

Morrison KD, Reekie EG, Jensen KIN (1998) Biocontrol of Common St. Johnswort *(Hypericum perforatum)* with *Chrysolina hyperici* and a Host-Specific *Colletotrichum gloeosporioides.* Weed Technol 12:426–435

Moss SR (1985) The influence of crop variety and seed rate on *Alopecurus yosuroides* competition in winter cereals. In: Proceedings of Brighton crop protection conference – weeds, pp 701–708

Muller E, Jud P, Nentwig W (2011) Artificial infection of *Cirsium arvense* with the rust pathogen *Puccinia punctiformis* by imitation of natural spore transfer by the weevil *Ceratapion onopordi*. Weed Res 51:209–213

Nalewaja JD (1978) Weed control in cereals—now and in the future. In: Parsons WT, Eady FC, Richardson RG (eds) Proceedings of the 1st conference of the council of Australian weed science societies. council of Australian weed science societies, Melbourne, pp 215–222

Nelson AG, Pswarayi A, Quideau S, Frick B, Spaner D (2012) Yield and weed suppression of crop mixtures in organic and conventional systems of the western Canadian prairie. Agron J 104:756–762

Nelson WA, Kahn BA, Roberts BW (1991) Screening cover crops for use in conservation tillage systems for vegetables following spring plowing. HortScience 26:860–862

O'Donovan JT, Blackshaw RE, Harker KN, Clayton GW (2006) Wheat seeding rate influences herbicide performance in wild oat (*Avena fatua* L.). Agron J 98:815–822

Odonovan JT, Destremy EA, Osullivan PA, Dew DA, Sharma AK (1985) Influence of the relative-time of emergence of wild oat (*Avena atua)* on yield loss of barley (*Hrdeum ulgare*) and wheat (*Titicum aestivum*). Weed Sci 33:498–503

Olsen J, Kristensen L, Weiner J (2006) Influence of sowing density and spatial pattern of spring wheat (*Triticum aestivum*) on the suppression of different weed species. Weed Biol Manag 6:165–173

Ortiz Monasterio JI, Sayre KD, Rajaram S, McMahon M (1997) Genetic progress in wheat yield and nitrogen use efficiency under four nitrogen rates. Crop Sci 37:898–904

Pavlychenko TK, Harrington JB (1934) Competitive efficiency of weeds and cereal crops. Can J Res 10:77–94

Paynter BH, Hills AL (2009) Barley and Rigid Ryegrass *(Lolium rigidum)* Competition is Influenced by Crop Cultivar and Density. Weed Technol 23:40–48

Peltzer S (1999) Increased crop density reduces weed seed production without increasing screenings. In: Bishop AC, Boersma M, Barnes CD (eds) Weed management into the 21st century: do we know where we're going?. 12th Australian Weeds Conference, Papers and Proceedings, Hobart, Tasmania, Australia, 12–16 September 1999, pp 510–512

Pridham JC, Entz MH, Martin RC, Hucl RJ (2007) Weed, disease and grain yield effects of cultivar mixtures in organically managed spring wheat. Can J Plant Sci 87:855–859

Putnam AR, Defrank J, Barnes JP (1983) Exploitation of allelopathy for weedcontrol in annual and perennial cropping systems. J Chem Ecol 9:1001–1010

Radford BF, Wilson BF, Cartledge O, Watkins FB (1980) Effect of wheat seeding rate on wild oat competition. Aust J Exp Agric 20:77–81

Rasmussen K, Rasmussen J, Petersen J (1996) Effects of fertilizer placement on weeds in weed harrowed spring barley. Acta Agri Scand Sect B: Soil Plant Sci 46:192–196

Rebetzke GJ, Richards RA (1999) Genetic improvement of early vigour in wheat. Aust J Agric Res 50:291–301

Rebetzke G, Appels R, Morrison A, Richards R, McDonald G, Ellis M, Spielmeyer W, Bonnett D (2001a) Quantitative trait loci on chromosome 4B for coleoptile length and early vigour in wheat (*Triticum aestivum* L.). Crop Past Sci 52:1221–1234

Rebetzke GJ, Appels R, Morrison AD, Richards RA, McDonald G, Ellis MH, Spielmeyer W, Bonnett DG (2001b) Quantitative trait loci on chromosome 4B for coleoptile length and early vigour in wheat (*Triticum aestivum* L.). Aust J Agric Res 52:1221–1234

Rebetzke G, Ellis M, Bonnett D, Richards R (2007) Molecular mapping of genes for coleoptile growth in bread wheat (*Triticum aestivum* L.). Theor Appl Genet 114:1173–1183

Reid T, Yang R-C, Salmon D, Spaner D (2009a) Should spring wheat breeding for organically managed systems be conducted on organically managed land? Euphytica 169:239–252

Reid TA, Navabi A, Cahill JC, Salmon D, Spaner D (2009b) A genetic analysis of weed competitive ability in spring wheat. Can J Plant Sci 89:591–599

Reid TA, Yang RC, Salmon DF, Navabi A, Spaner D (2011) Realized gains from selection for spring wheat grain yield are different in conventional and organically managed systems. Euphytica 177:253–266

Richards RA (1983) Manipulation of leaf-area and its effect on grain-yield in droughted wheat. Aust J Agric Res 34:23–31

Richards RA, Lukacs Z (2002) Seedling vigour in wheat-sources of variation for genetic and agronomic improvement. Aust J Agric Res 53:41–50

Roberts HA, Feast PM (1972) Fate of seeds of some annual weeds in different depths of cultivated and undisturbed soil. Weed Res 12:316–324

Rodriguez EE (2006) Effect of cultivar mixture on the competitive ability of barley against weeds. Institutionen For vaxtproduktionsekologi Sveriges Lantbruksuniversitet

Röder MS, Huang XQ, Ganal MW (2005) Wheat microsatellites: potential and implications. In: Lörz H, Wenzel G (eds) Molecular marker systems in plant breeding and crop improvement. Springer, Heidelberg, pp 255–266

Ross SM, King JR, Izaurralde RC, O'Donovan JT (2001) Weed suppression by seven clover species. Agron J 93:820–827

Salvi S, Tuberosa R (2005) To clone or not to clone plant QTLs: present and future challenges. Trends Plant Sci 10:297–304

Sandhu RS, Gill BS, Chahal PS (2010) Competitive ability of wheat (*Triticum aestivum* L.) against wild oats (*Avena ludoviciana* Dur.) as influenced by date of sowing, seed rate and spacing. Indian J Ecol 37:97–100

Sarrantonio M, Gallandt E (2003) The role of cover crops in North American cropping systems. J Crop Prod 8:53–74

Schaedler CE, Fleck NG, Ferreira FB, Lazaroto CA, Rizzardi MA (2009) Morphological traits in oat plants cultivars as indicators of competitive potential against weeds. Ciencia Rural 39:1313–1319

Seavers GP, Wright KJ (1999) Crop canopy development and structure influence weed suppression. Weed Res 39:319–328

Smil V (2000) Feeding the world: a challenge for the twenty-first century. The MIT Press, Cambridge

Spielmeyer W, Hyles J, Joaquim P, Azanza F, Bonnett D, Ellis ME, Moore C, Richards RA (2007) A QTL on chromosome 6A in bread wheat (*Triticum aestivum*) is associated with longer coleoptiles, greater seedling vigour and final plant height. Theor Appl Genet 115:59–66

Stute JK, Posner JL (1995) Synchrony between legume nitrogen release and corn demand in the upper Midwest. Agron J 87:1063–1069

Tanksley SD, McCouch SR (1997) Seed banks and molecular maps: unlocking genetic potential from the wild. Science 277:1063–1066

Taylorson RB (1987) Environmental and chemical manipulation of weed seed dormancy. Rev Weed Sci 3:135–154

Teasdale JR, Coffman CB, Mangum RW (2007) Potential long-term benefits of no-tillage and organic cropping systems for grain production and soil improvement. Agron J 99:1297–1305

ter Steege MW, den Ouden FM, Lambers H, Stam P, Peeters AJM (2005) Genetic and physiological architecture of early Vigor in *Aegilops tauschii*, the D-genome donor of hexaploid wheat. A quantitative trait loci analysis. Plant Physiol 139:1078–1094

Trethowan RM, Mahmood T, Ali Z, Oldach K, Garcia AG (2012) Breeding wheat cultivars better adapted to conservation agriculture. Field Crops Res 132:76–83 van Bueren ETL, Goldringer I, Østergard H (2005) In: Proceedings of the COST SUSVAR/ECO-PB workshop on organic plant breeding strategies and the use of molecular markers, Louis Bolk Institute, Driebergen, 17–19 January 2005

van Bueren ETL, Backes G, de Vriend H, Ostergard H (2010) The role of molecular markers and marker assisted selection in breeding for organic agriculture. Euphytica 175:51–64

van Bueren ETL, Jones SS, Tamm L, Murphy KM, Myers JR, Leifert C, Messmer MM (2011) The need to breed crop varieties suitable for organic farming, using wheat, tomato and broccoli as examples: a review. Njas-Wageningen J Life Sci 58:193–205

Verschwele A, Niemann P, European Weed Res SOC (1993) Indirect weed-control by selection of wheat cultivars. In :Proceedings of 8th European weed research society symposium quantitative approaches in weed and herbicide research and their practical application (Ewrs 1993), vols 1 and 2, pp 799–806

Walker SR, Robinson GR, Medd RW (1998) Management of wild oats and paradoxa grass with reduced dependence on herbicides. In: Michalk DL, Pratley JA (eds) Proceedings of the 9th Australian agronomy conference. Australian society of agronomy: Wagga Wagga, pp 572–574

Weiner J, Griepentrog HW, Kristensen L (2001) Suppression of weeds by spring wheat *Triticum aestivum* increases with crop density and spatial uniformity. J Appl Ecol 38:784–790

Wicks GA, Ramsel RE, Nordquist PT, Schmidt JW, Challaiah (1986) Impact of wheat cultivars on establishment and suppression of summer annual weeds. Agron J 78:59–62

Wiese AF, Collier JW, Clark LE, Havelka UD (1964) Effect of weeds and cultural practices on sorghum yields. Weeds 12:209–211

Wissuwa M, Mazzola M, Picard C (2009) Novel approaches in plant breeding for rhizosphere-related traits. Plant Soil 321:409–430

Wojciechowski T, Gooding MJ, Ramsay L, Gregory PJ (2009) The effects of dwarfing genes on seedling root growth of wheat. J Exp Bot 60:2565–2573

Worthington M, Reberg-Horton C (2013) Breeding cereal crops for enhanced weed suppression: optimizing allelopathy and competitive ability. J Chem Ecol 39:213–231

Wu H, Pratley J, Ma W, Haig T (2003) Quantitative trait loci and molecular markers associated with wheat allelopathy. Theor Appl Genet 107:1477–1481

Yenish JR, Young FL (2004) Winter wheat competition against jointed goatgrass (*Aegilops cylindrica*) as influenced by wheat plant height, seeding rate, and seed size. Weed Sci 52:996–1001

Yue B, Xue W-Y, Luo L-J, Xing Y-Z (2006) QTL analysis for flag leaf characteristics and their relationships with yield and yield traits in rice. Acta Genetica Sin 33:824–832

Zhang K-P, Zhao L, Hai Y, Chen G-F, Tian J-C (2008) QTL mapping for adult-plant resistance to powdery mildew, lodging resistance, and internode length below spike in wheat. Acta Agron Sin 34:1350–1357

Zhang ZH, Yu SB, Yu T, Huang Z, Zhu YG (2005) Mapping quantitative trait loci (QTLs) for seedling-vicror using recombinant inbred lines of rice (Oryza sativa L.). Field Crops Research 91:161–170

Zhang ZH, Li P, Wang LX, Hu ZL, Zhu LH, Zhu YG (2004) Genetic dissection of the relationships of biomass production and partitioning with yield and yield related traits in rice. Plant Science 167:1–8

Zhao DL, Atlin GN, Bastiaans L, Spiertz JHJ (2006) Comparing rice germplasm groups for growth, grain yield and weed-suppressive ability under aerobic soil conditions. Weed Res 46:444–452

Zhou L, Wang J-K, Yi Q, Wang Y-Z, Zhu Y-G, Zhang Z-H (2007) Quantitative trait loci for seedling vigor in rice under field conditions. Field Crops Research 100:294–301

Zhu LH, Zhong DB, Xu JL, Yu SB, Li ZK (2008) Differential expression of lodging resistance related QTLs in rice (Oryza sativa L.). Plant Science 175:898–905

Zuo S, Liu G, Li M (2012) Genetic basis of allelopathic potential of winter wheat based on the perspective of quantitative trait locus. Field Crops Res 135:67–73

Chapter 4
Breeding Wheat for Organic Agriculture

Abstract Plant breeders around the globe emphasize on improving yield, adaptation, disease resistance and quality in conventional management systems where the use of synthetic fertilizers, pesticides and herbicides is routine and thus, precluding selection for traits conferring competitive ability. This Chapter focuses on breeding wheat cultivars for disease resistance, grain quality, allelopathy and earliness for organically managed lands. The combined effect of allelopathy and competitive ability can determine the competitiveness of a given crop species to achieve maximum weed suppression. Weed suppressive rice cultivars have been developed and released for commercial cultivation in USA and China, whereas breeding efforts are being done in various parts of the world including Canada to develop weed suppressive/competitive wheat and barley ideotypes.

Keywords Allelopathy · Baking quality · Barley · Disease resistance · Early maturity · Rice · Wheat

4.1 Breeding for Disease Resistance

Disease resistance is a major issue for both conventional and organic management systems but plant health in organic systems is a greater challenge due to the avoidance of pesticides and insecticides. Bunts, smuts, Septoria, and Fusarium Head Blight (FHB) have been reported to be the most recognized diseases of wheat in organic agriculture (Engelke 1992). Diseases such as rusts and powdery mildew, which are influenced by nitrogen application, planting date, and crop density, are less important in organic systems because of their occurrence in the later stages of crop growth (Kunts 1983) particularly due to severe weed competition during early stages of crop establishment/development. However, Cauvain (2003) considered stripe rust as an integral part of wheat breeding for organic agriculture. In a comparison study between organic and conventional management systems, Neacsu et al. (2010) reported less frequency and incidence of stripe rust in organic systems

M. Asif et al., *Managing and Breeding Wheat for Organic Systems*,
SpringerBriefs in Agriculture, DOI: 10.1007/978-3-319-05002-7_4,
© The Author(s) 2014

and a similar finding for leaf rust has also been reported, when cow manure has been used for fertilizer (Rodgers-Gray and Shaw 2000). Soil (*Septoria tritici* blotch and *Drechslera tritici repentis*) and seed (smuts and bunts) borne diseases cannot be controlled by cultural measures. There are no efficient and practical seed dressing treatments available to control seed borne diseases in organic agriculture. Thus, breeding-resistant cultivars are the only choice available for these pests. Furthermore, the level of resistance should not be considered the sole selection criteria to develop/breed wheat cultivars. The ability of crop plants to produce optimum/maximum grain yield along with better quality regardless of the pressure of infectious diseases under field conditions must be given top priority during selection (Lammerts 2002). Therefore, resistant cultivars/lines should not be only resistant, but morphological/phenotypic traits must guarantee its resistance during high disease pressures. Such selection criteria is not a priority while selecting superior wheat lines against diseases under conventional management systems. For example, Kunts (1983) reported that occurrence of *S. nodorum* can be influenced by plant architecture. Greater distance between flag leaf and spike can reduce the transfer of spores of *S. nodorum* from leaves to spike by rain drops (Kunts 1983) and can also minimize infection of spike in case of attack by *Fusarium* spp. (Engelke 1992).

4.2 Breeding for Quality

Baking/bread quality of wheat is the capacity of flour to guarantee a high volume and elastic bread with uniform pores (Neacsu et al. 2010) and is influenced by a number of factors including color, texture, nutritional value, and flavor (Cauvain 2003). Bread making quality is directly related to the retention of CO_2 during the fermentation process, water absorption capacity, and dough mixing behavior. Grain protein content is one of the major determinants of wheat bread making quality (Ohm and Chung 1999) because it provides dough strength and allows the retention of CO_2 during fermentation (Gooding et al. 1999). Grain protein content ranging from 10.5 to 13.5 % is desirable for bread making, whereas wheat cultivars having grain protein content less than 10 % are used to make cookies, cakes, or crackers or are blended with grains having protein content greater than 14 % (Mason et al. 2007). Grain protein content is the most widely studied trait in bread wheat, and it is affected by genetics, environmental conditions, and management practices (Horvat et al. 2009; Mason et al. 2007; Stanciu and Neacsu 2008). Research studies on bread making quality of wheat under organically managed lands are limited in number and sometimes contradictory. Annett et al. (2007) reported higher levels of whole meal protein in grains of organically produced wheat than conventional grains; however, protein content of grains in both systems was greater than 14 %. In a long-term (21 years) study, Maeder et al. (2007) reported no difference in nutritional value (amino acid compositions, protein content, mineral contents, and trace element contents) and baking quality of wheat grains in organic and conventional systems. Similar findings in terms of grain protein content (Shier et al. 1984) and nitrogen concentration (Ryan et al. 2004) have also been documented. The environment was reported as the major driver of variation in protein content (Fowler and Delaroche 1975) along with nitrogen fertilizer and soil moisture content (Preston et al. 2001; Shier et al. 1984). With the application of nitrogen fertilizer, an increase in grain protein content and gluten strength has been reported in various studies (Ames et al. 2003; Gooding et al. 1993; Johansson et al. 2003; Lerner et al. 2006). Mazzoncini et al. (2007) reported 20 % less protein content of grain samples from organic than conventionally managed systems. They further reported poor bread making quality of organic samples; however, there were no visual differences for crumb volume and crust thickness. Similar findings of higher grain protein content of wheat grown in conventional management system have also been reported in various other studies (Baeckstrom et al. 2004; Starling and Richards 1993).

4.3 Breeding for Allelopathy

Many crop plants produce and release certain chemicals that are toxic to other plants growing in the same vicinity. The phenomenon of producing such toxic chemicals is referred to as "allelopathy". These toxic chemicals/allelochemicals are also released by microbes during the process of residue decomposition. Allelochemicals either originating from crop plants or microbes can have direct effects on weed germination, growth, and establishment. Allelochemicals from weeds can also have adverse effect on the crop plants. Allelopathy is gaining interest among cereal breeders and weed scientists as a tool of integrated weed management, especially for organic and low-input farming systems (Belz 2007; Olofsdotter et al. 2002). It has been suggested that an allelopathic effect of wheat can have a wider spectrum to control many weed species (Worthington and Reberg-Horton 2013). Bertholdsson (2005) screened wheat and barley genotypes for allelopathic potential using an agar bioassay to control rye grass (*Lolium perenne*). The study concluded that early crop biomass and allelopathic activity were the only studied traits which contributed significantly to crop competitiveness. The author suggested that an increase in the allelopathic effect of wheat up to a level of many highly allelopathic barley cultivars can result in improved weed competitiveness by at least 60 %. The success rate will be highly dependent on the availability of genetic resources, screening methods, number of genes involved, and the heritability of the allelopathic traits. Bertholdsson (2011) reported greater allopathic activity of winter wheat cultivars against mustard (*Sinapis alba*) than ryegrass (*Lolium perenne*), suggesting the importance of allelopathy to control locally and economically important weed species. Wu et al. (2000) found that major genes are involved in controlling seedling allelopathy in wheat, and observed significant differences for seedling allelopathy among 453 wheat cultivars from 50 countries against annual rye grass (*Lolium rigidum* L.).

Bertholdsson (2010) screened 813 spring wheat cultivars (525 Swedish cultivars/ lines with the rest from eastern and central Europe, North and South America, Asia, and Africa) for allelopathic activity using an agar-based bioassay. Based on screening, two cultivars, *i.e.*, Mohan 73 (exhibiting high allelopathic activity) and Zebra (having low allelopathic activity) were identified and crossed to perform agronomic and allelopathic selections over various segregating generations. Their study concluded that lines exhibiting high allelopathic activity were able to reduce weed biomass by at least 19 %, thereby enhancing competitive ability in wheat against perennial ryegrass (*Lolium perenne*). Wu et al. (2003) identified two major QTLs on chromosome 2B for wheat allelopathic activity using a DH population derived from a cross between the Australian cultivars Tasman (strong allelopathic) and Sunco (moderate allelopathic). This study suggested that breeding efficiency for competitive ability can be enhanced using marker assisted selection (MAS) for allelopathic activity. Genes and QTLs (on chromosomes 4A, 4B, 4D, and 5B) controlling hydroxamic acid (allelochemicals) accumulation were identified and mapped in wheat using a mapping population developed using substitution lines from Chinese spring cultivar with high hydroxamic acid and Cheyenne, with low hydroxamic acid (Niemeyer and Jerez 1997).

Allelopathic potential of *Triticum*, *Secale*, *Triticosecale* species along with substitution and translocation lines (wheat–rye) were investigated to enhance competitive ability against weeds where the authors identified various substitution and translocation wheat–rye lines having high allelopathic activity. These identified lines with substitutions for 1R or 2R exhibited maximum weed suppression/ allelopathic activity (Bertholdsson et al. 2012). Allelochemicals like Benzoxazinoids (BX) were reported to suppress weeds, pests, and diseases and be present in wheat and rye (Carlsen et al. 2009). In wheat, BX biosynthesis starts right after germination and reaches its maximum within 7–10 days (Argandona et al. 1981), whereas in rye it retained and reached a maximum after 60 days following germination (Burgos et al. 1999). Therefore, it was recommended that genes responsible for BX biosynthesis in rye can be manipulated/used in wheat to prolong its allelopathic activity (Bertholdsson et al. 2012). Root exudates and crop residues are potential means by which wheat cultivars can reduce weed growth and establishment. Various studies have shown considerable genetic variation for root exudates like benzoic, para-coumaric acid, vanillic, and ferulic acid that exist in spring wheat, barley, and oat which can be exploited to enhance competitiveness (Baghestani et al. 1999; Wu et al. 1999). Rao and Pandya (1992) found that root exudates of wheat plants release an untriacontane compound that had an inhibitory effect on *Asphodelus tenuifolius* (wild onion) growth up to 60 days after wheat germination. Similarly, a cold aqueous extract of wheat straw reduced growth and germination of annual broad leaf weed species including sesbania, hemp, prickly sida, ivy leaf, velvet leaf, and morning glory (Blum et al. 1991). In a similar study, Lovett (1983) reported a 53 % reduction in weed populations as a result of wheat residues. A severe reduction in prickly sida (teaweed) and morning glory occurred using wheat mulch and ferulic acid (as a wheat extract) in the field studies (Liebl and Worsham 1983, 1987). Therefore, it may be worthwhile to retain and improve allopathic effects of wheat to enhance competitive ability against weeds by considering and selecting allelopathic traits in current and future wheat breeding programs.

4.4 Breeding for Early Maturity

Early maturity is an important global breeding objective in spring wheat. Early maturity ensures early crop harvest and may also assist wheat escape from various diseases, heat, and drought stress (Sleper and Poehlman 2006). In high northern latitudes such as western Canada, earliness is required due to the short growing season and pre-harvest sprouting of physiologically mature grains on the plants. Wheat breeders need to consider the frost-free period when selecting for these regions because frost both at anthesis and later in the season can significantly reduce production and quality. In addition, early maturing cultivars need to be high yielding in order to be competitive with other crops.

One of the biggest challenges for organic agricultural systems is the strong competition from weeds. One of the solutions to this problem is delayed planting. This will allow enough time for the weeds to germinate and then for the farmer to cultivate the land and uproot the weeds. Planting late on such field greatly reduces the incidence of weeds. However, this strategy requires availability of varieties with early maturity so as to fit into the growing season, especially in the short growing season prevalent in the higher northern latitudes. Early maturing varieties will require relatively less time to complete their life cycle and due to less competition from weeds it will likely yield relatively higher in organic systems.

Growth and the various developmental phases of wheat crop are controlled by vernalization (*Vrn*) response, photoperiod (*Ppd*) sensitive, and genes conferring earliness *per se* (Košner and Pánková 1998). These three systems of genes and the interactions of these genes with temperatures during growth period (Gororo et al. 2001) play a vital role in the adaptation of wheat and its yield potential in a diverse range of environments, including organically managed lands. Photoperiod and vernalization genes act to accelerate or delay flowering in response to seasonal changes in the environment, to ensure that flowering occurs at optimum temperatures (Law and Worland 1997). At high northern latitudes, for instance, vernalization genes delay ear initiation in winter wheat to protect the delicate floral organs from damage due to extreme low temperatures. Similarly, in regions of the world where summers are very hot to grow wheat, photoperiod insensitive genes accelerate ear initiation and development with increasing day length during late winter and early spring, ensuring the completion of the reproductive phase before the onset of high temperatures. The *Vrn* gene system accounts for about 70–75 %, the *Ppd* gene system for about 20–25 %, and earliness *per se* for about 5 % of the genetic variability in the heading date of bread wheat (Stelmakh 1998). Genes influencing flowering in wheat are distributed over almost all chromosomes (Law and Worland 1997), and genes conferring vernalization and photoperiod responses, and earliness *per se* appear to be located on each of the three homoeologous chromosomes of a group (Worland and Snape 2001).

Among the known vernalization response genes, *Vrn-A1* inhibits vernalization requirement the most, followed by *Vrn-D1*, *Vrn4*, and *Vrn-B1*, respectively (Goncharov 2004). This implies that, other genetic factors remaining constant, plants with dominant *Vrn-A1* will head first, whereas those having dominant *Vrn-B1* will head last. Shindo et al. (2003), however, reported that the strong spring habit of *Vrn-B1* is equivalent to *Vrn-A1*. Existence of multiple alleles for all of the dominant *Vrn* genes has been reported (Klaimi and Qualset 1974; Kuspira et al. 1986; Law et al. 1976). Besides being the strongest inhibitor of vernalization requirement, *Vrn-A1* also masks the effect of dominant *Vrn-B1* and *Vrn-D1* alleles (Pugsley 1971). Roberts and Larson (1985), however, reported that *Vrn-A1* is neither always fully dominant nor always epistatic. Stelmakh (1993) studied the effects of *Vrn* genes on heading time and other agronomic traits in three sets of isogenic lines of bread wheat, and reported that *Vrn* genes act as non-complementary genes with classical (*Vrn-A1*) or incomplete (*Vrn-D1*) dominant epistasis.

Yan et al. (2004) developed Polymerase chain reaction (PCR)-based DNA markers specific for vernalization alleles *Vrn-A1a* and *Vrn-A1b* on the basis of deletions or insertions in the *Vrn-A1* promoter. They used these markers to screen 132 spring wheat genotypes from Canada, USA, and Argentina, and found *Vrn-A1a* in 55 % and Vrn-A1b in 6 % genotypes. Zhang et al. (2008) also found high proportion of *Vrn-A1a* allele in spring cultivars from China. Fu et al. (2005) developed allele-specific DNA markers for the alleles *vrn-A1*, *Vrn-A1c*, *vrn-B1*, *Vrn-B1*, *vrn-D1*, and *Vrn-D1*. These markers were based on the absence or presence of large deletions within intron 1. The newly developed markers were used to characterize 117 spring wheat genotypes from California and Argentina for allelic composition at Vrn-1 loci. *Vrn-A1a/b* was the most frequent allele, present either singly or in combination with other *Vrn-1* alleles, followed by alleles *Vrn-B1* and *Vrn-D1*. They did not find Vrn-A1c in the studied genotypes. Iqbal et al. (2007b) found *Vrn-A1a* allele in 85 % Canadian spring wheat varieties/lines, followed by *Vrn-B1* in 50 % wheat varieties/lines, and *Vrn-A1b* in one genotype. *Vrn-A1c* and *Vrn-D1* were absent from the 40 genotypes studied. Spring growth habit allele *Vrn-A1a* was identified in 36 % Pakistani wheat cultivars either singly or in combination with spring habit *Vrn-B1* and *Vrn-D1* alleles (Iqbal et al. 2011). They found dominant *Vrn-A1c* allele in two wheat cultivars but did not find *Vrn-A1b* in the studied cultivars. Spring habit *Vrn-B1* was found in the highest frequency (64 %) either singly or with spring habit alleles *Vrn-A1a*, *Vrn-A1c,* and *Vrn-D1*. Spring habit *Vrn-D1* was identified in 61 % Pakistani cultivars. Dominant allele of *Vrn-D1* was singly found in 25 % cultivars and along with *Vrn-B1* in 29 % cultivars. Dominant *Vrn-B3* was absent in all cultivars studied. It was also reported that the three major Vrn genes have differential effect on flowering time, plant height, and grain yield components (Stelmakh 1993; Stelmakh 1998). Genotypes having two spring habit alleles in combination matured early and had higher grain yield. Wheat genotypes with three spring habit alleles were found to be early maturing but low yielding. These findings suggest that specific dominant genes in spring wheat can be combined to develop early maturing cultivars with relatively higher grain yield potential. Incorporation of *Vrn-D1* has been recommended in spring wheat breeding programs.

Photoperiod response is the second important genetic factor that determines flowering time, and therefore, the wheat's adaptation to diverse climatic conditions. As mentioned previously, vernalization is the primary factor determining winter and spring growth habit. However, flowering time of autumn sown spring or winter wheat varieties is not significantly affected by the presence of different *Vrn* genes, as their vernalization requirement is generally fulfilled (Worland and Snape 2001). Under such conditions, flowering time is determined mainly by sensitivity/insensitivity to photoperiod (day length). A photoperiod insensitive wheat variety can immediately switch to reproductive growth when temperature rises in the spring, whereas a photoperiod sensitive wheat variety remains vegetative until the day length increases and its photoperiod requirement is satisfied (Worland and Snape 2001). If the photoperiod requirement of sensitive varieties is not fulfilled, flowering is delayed. The extent of lateness in such a situation depends on which photoperiod response genes is present in the variety and the latitude of the specific growing region. In North America, more photoperiod insensitive wheat cultivars are being developed because breeders are increasingly planting winter nurseries in southern latitudes to obtain an extra breeding generation (Dyck et al. 2004). However, photoperiod sensitivity has been found to be advantageous in the North American higher latitudes with respect to local adaptation, yield stability, and high productivity (Busch et al. 1984; Dyck et al. 2004). This area needs further investigation before making general recommendations regarding the utilization of particular photoperiod response mechanisms in wheat breeding programs in these regions.

Photoperiodic and vernalization response genes affect flowering time of wheat under specific day length and growing temperature, whereas genes of earliness *per se* control heading time independent of environmental stimuli (Worland 1996). Because of their major effects on heading time, vernalization and photoperiod response genes have been studied in detail, while earliness *per se* genes have not been fully investigated (Kato and Wada 1999). Major vernalization and photoperiod genes may be regarded as "modifiers of earliness" because they influence flowering only in response to certain environmental conditions, but earliness is determined by a minimum vegetative growth that can initiate floral primordia independent of external stimuli (Kato and Wada 1999). Earliness *per se* is a quantitatively inherited trait that is controlled by several minor genes whose effects can be detected only in the absence of the confounding effects of photoperiod and vernalization response genes (Kato and Wada 1999). Earliness *per se* is highly heritable and can, therefore, be effectively utilized in breeding programs to shorten wheat's life cycle independent of other environmental factors known to modify flowering time (Kato and Wada 1999).

In the genetic analyses of flowering time, earliness *per se* genes are found having relatively smaller effects and are generally mapped as QTL rather than as major genes (Kato et al. 1999). This is partly because most of the earliness *per se*

QTL have been identified in mapping populations developed for locating major vernalization or photoperiod response genes (Kato et al. 1999). Detailed study of earliness *per se* genes require specific genetic material and environmental conditions, where the presence of such genes is not obscured by the effects of *Ppd* and *Vrn* genes. Earliness *per se* genes have been found to influence flowering time of wheat in many studies but none of these reported that *eps* genes have pleiotropic effects on any other traits of agronomic importance (Worland and Snape 2001). However, due to their independent nature, such genes could be of value to wheat breeders in modifying flowering time, further adding to the wide adaptability of wheat.

Besides influencing flowering time, genes of all three systems also seem to have pleiotropic effects on other aspects of plant's growth and development with respect to grain yield potential. The geographical distribution of photoperiod and vernalization genes has demonstrated that these genes have different breeding values for a given set of environmental conditions. The characterization of individual genes and the identification of diagnostic markers for the key photoperiod, vernalization, and earliness *per se* genes may make it possible to "fine-tune" flowering time of wheat, further adding to the wide adaptability of this crop. The genetic basis of earliness has been extensively studied in western Canada (Iqbal et al. 2007b; Kamran et al. 2013) and elsewhere (Iqbal et al. 2011; Santra et al. 2009; Zhang et al. 2008). Iqbal et al. (2007a, b) demonstrated that wheat genotypes with dominant Vrn-A1 allele either singly or in combination with other Vrn alleles are early flowering, whereas those with dominant allele of Vrn-B1 and/or Vrn-D1 flower late. This suggests that wheat varieties with dominant allele of Vrn-A1 should be planted on organic lands as these genotypes can complete their life cycle within the short of window of organic farming system.

Grain yield, protein concentration, and early maturity are negatively correlated, posing a challenge to wheat breeders developing early maturing, high yielding, and high quality spring wheat (DePauw et al. 1995). The main determinants of time to maturity are time to anthesis and grain fill duration (GFD) (the period from anthesis to physiological maturity (Duguid and Brûlé-Babel 1994). As GFD and grain fill rate (GFR) determine final grain weight in wheat, a better understanding of the processes and factors governing maturity and grain filling may aid in decreasing time to maturity without reducing grain yield (Duguid and Brûlé-Babel 1994). Genetic variation for the rate and duration of the grain filling period has been reported in spring wheat (Bruckner and Frohberg 1987; Darroch and Baker 1995; Iqbal et al. 2007a; Talbert et al. 2001). There are inconsistent reports about the associations between GFD and grain yield. Some studies have indicated a positive association between GFD and grain weight and/or yield (Gebeyehou et al. 1982; Iqbal et al. 2007a; Sharma 1994; Sofield et al. 1977), while others reported no association between the two traits (Bruckner and Frohberg 1987; Nass and Reiser 1975; Talbert et al. 2001; Wang et al. 2002). The association between GFR and grain weight has been reported to be positive and consistent across different experiments (Bruckner and Frohberg 1987; Darroch and Baker 1995; Hunt et al. 1991; Iqbal et al. 2007a).

Keeping the above facts in perspective, organic breeders need to develop early maturing cultivars by shortening vegetative period without effecting GFD to a greater extent. This is likely to result in varieties with early maturity but no significant decrease in grain yield.

References

Ames NP, Clarke JM, Dexter JE, Woods SM, Selles F, Marchylo B (2003) Effects of nitrogen fertilizer on protein quantity and gluten strength parameters in durum wheat (*Triticum turgidum* L. var. durum) cultivars of variable gluten strength. Cereal Chem 80:203–211

Annett LE, Spaner D, Wismer WV (2007) Sensory profiles of bread made from paired samples of organic and conventionally grown wheat grain. J Food Sci 72:S254–S260

Argandona VH, Niemeyer HM, Corcuera LJ (1981) Effect of content and distribution of hydroxamic acids in wheat on infestation by the aphid *Schizaphis graminum*. Phytochemistry 20:673–676

Baeckstrom GL, Hanell U, Svensson G (2004) Baking quality of winter wheat grown in different cultivating systems, 1992–2001: A holistic approach. J Sustain Agri 24:53–79

Baghestani A, Lemieux C, Leroux GD, Baziramakenga R, Simard RR (1999) Determination of allelochemicars in spring cereal cultivars of different competitiveness. Weed Sci 47:498–504

Belz RG (2007) Allelopathy in crop/weed interactions: an update. Pest Manag Sci 63:308–326

Bertholdsson N-O, Andersson SC, Merker A (2012) Allelopathic potential of *Triticum* spp., *Secale* spp. and *Triticosecale* spp. and use of chromosome substitutions and translocations to improve weed suppression ability in winter wheat. Plant Breeding 131:75–80

Bertholdsson NO (2005) Early vigour and allelopathy: Two useful traits for enhanced barley and wheat competitiveness against weeds. Weed Res 45:94–102

Bertholdsson NO (2010) Breeding spring wheat for improved allelopathic potential. Weed Res 50:49–57

Bertholdsson NO (2011) Use of multivariate statistics to separate allelopathic and competitive factors influencing weed suppression ability in winter wheat. Weed Res 51:273–283

Blum U, Wentworth TR, Klein K, Worsham AD, King LD, Gerig TM, Lyu SW (1991) Phenolic-acid content of soils from wheat-no till, wheat-conventional till, and fallow-conventional till soybean cropping systems. J Chem Ecol 17:1045–1068

Bruckner P, Frohberg R (1987) Rate and duration of grain fill in spring wheat. Crop Sci 27:451–455

Burgos NR, Talbert RE, Mattice JD (1999) Cultivar and age differences in the production of allelochemicals by *Secale cereale*. Weed Sci 47:481–485

Busch R, Elsayed F, Heiner R (1984) Effect of daylength insensitivity on agronomic traits and gain protein in hard spring wheat. Crop Sci 24:1106–1109

Carlsen SCK, Kudsk P, Laursen B, Mathiassen SK, Mortensen AG, Fomsgaard IS (2009) Allelochemicals in rye *(Secale cereale* L.): cultivar and tissue differences in the production of benzoxazinoids and phenolic acids. Nat Prod Commun 4:199–208

Cauvain SP (2003) Wheat and its special properties. In: Cauvain XP (ed) Bread making: improving quality. Woodhead Publishing Ltd, Cambridge, pp 2–3

Darroch B, Baker R (1995) Two measures of grain filling in spring wheat. Crop Sci 35:164–168

DePauw R, Boughton G, Knott D (1995) Hard red spring wheat. Harvest of gold: the history of field crop breeding in Canada University Extension Press, University of Saskatchewan, Saskatoon, pp 5–35

Duguid S, Brûlé-Babel A (1994) Rate and duration of grain filling in five spring wheat (*Triticum aestivum* L.) genotypes. Can J Plant Sci 74:681–686

Dyck J, Matus-Cadiz M, Hucl P, Talbert L, Hunt T, Dubuc J, Nass H, Clayton G, Dobb J, Quick J (2004) Agronomic performance of hard red spring wheat isolines sensitive and insensitive to photoperiod. Crop Sci 44:1976–1981

Engelke F (1992) Ertrag und Ertragsbildung von Winterweizen, Winterrogen und Winteriticale im Organischen Landbau-Aswertung von Sortenversuchen in drei Versuchenjahren. Faculty of Agriculture. University of Bonn

Fowler DB, Delaroche IA (1975) Wheat quality evaluation.3. Influence of genotype and environment. Can J Plant Sci 55:263–269

Fu D, Szűcs P, Yan L, Helguera M, Skinner JS, von Zitzewitz J, Hayes PM, Dubcovsky J (2005) Large deletions within the first intron in VRN-1 are associated with spring growth habit in barley and wheat. Mol Genet Genomics 273:54–65

Gebeyehou G, Knott D, Baker R (1982) Relationships among durations of vegetative and grain filling phases, yield components, and grain yield in durum wheat cultivars. Crop Sci 22:287–290

Goncharov N (2004) Response to vernalization in wheat: its quantitative or qualitative nature. Cereal Res Commun 32:323–330

Gooding MJ, Cannon ND, Thompson AJ, Davies WP (1999) Quality and value of organic grain from contrasting breadmaking wheat varieties and near isogenic lines differing in dwarfing genes. Biol Agric Hortic 16:335–350

Gooding MJ, Davies WP, Thompson AJ, Smith SP (1993) The challenge of achieving breadmaking quality in organic and low input wheat in the UK: a review. Aspects of Appl Biol 36:189–198

Gororo N, Flood R, Eastwood R, Eagles HA (2001) Photoperiod and vernalization responses in Triticum turgidum × T. tauschii synthetic hexaploid wheats. Ann Bot 88:947–952

Horvat D, Drezner G, Magdic D, Simic G, Dvojkovic K, Lukinac J (2009) Effect of an oxidizing improver on dough rheological properties and bread crumb structure in winter wheat cultivars (Triticum aestivum L.) with different gluten strength. Rom Agri Res 26:35–40

Hunt L, Gvd Poorten, Pararajasingham S (1991) Postanthesis temperature effects on duration and rate of grain filling in some winter and spring wheats. Can J Plant Sci 71:609–617

Iqbal M, Navabi A, Salmon D, Yang RC, Spaner D (2007a) Simultaneous selection for early maturity, increased grain yield and elevated grain protein content in spring wheat. Plant Breeding 126:244–250

Iqbal M, Navabi A, Yang R-C, Salmon DF, Spaner D (2007b) Molecular characterization of vernalization response genes in Canadian spring wheat. Genome 50:511–516

Iqbal M, Shahzad A, Ahmed I (2011) Allelic variation at the Vrn-A1, Vrn-B1, Vrn-D1, Vrn-B3 and Ppd-D1a loci of Pakistani spring wheat cultivars. Electron J Biotechnol 14:1–2

Johansson E, Prieto-Linde ML, Svensson G, Jonsson JO (2003) Influences of cultivar, cultivation year and fertilizer rate on amount of protein groups and amount and size distribution of mono- and polymeric proteins in wheat. J Agric Sci 140:275–284

Kamran A, Randhawa HS, Pozniak C, Spaner D (2013) Phenotypic effects of the flowering gene complex in Canadian spring wheat germplasm. Crop Sci 53:84–94

Kato K, Miura H, Sawada S (1999) Detection of an earliness per se quantitative trait locus in the proximal region of wheat chromosome 5AL. Plant Breeding 118:391–394

Kato K, Wada T (1999) Genetic analysis and selection experiment for narrow-sense earliness in wheat by using segregating hybrid progenies. Breeding Sci 49:233–238

Klaimi Y, Qualset CO (1974) Genetics of heading time in wheat (Triticum aestivum L.). II. The inheritance of vernalization response. Genetics 76:119–133

Košner J, Pánková K (1998) The detection of allelic variants at the recessive vrn loci of winter wheat. Euphytica 101:9–16

Kunts P (1983) Entwicklungsstufen bei Gerste und Weizen - ein Beitrag zu einem Leitbild für die Züchtung. Naturwissenschaft 39:23–37

Kuspira J, Maclagan J, Kerby K, Bhambhani R (1986) Genetic and cytogenetic analyses of the A genome of Triticum monococcum. II. The mode of inheritance of spring versus winter growth habit. Can J Genet Cytol 28:88–95

Lammerts VBET (2002) Organic plant breeding and propagation: concepts and strategies. Wageningen University, The Netherland

Law C, Worland A (1997) Genetic analysis of some flowering time and adaptive traits in wheat. New Phytol 137:19–28

Law C, Worland A, Giorgi B (1976) The genetic control of ear-emergence time by chromosomes 5A and 5D of wheat. Heredity 36:49–58

Lerner SE, Seghezzo ML, Molfese ER, Ponzio NR, Cogliatti M, Rogers WJ (2006) N- and S-fertiliser effects on grain composition, industrial quality and end-use in durum wheat. J Cereal Sci 44:2–11

Liebl RA, Worsham AD (1983) Inhibition of pitted morning glory (*Ipomoea lacunosa* L) and certain other weed species by phytotoxic components of wheat (*Triticum aestivum* L) straw. J Chem Ecol 9:1027–1043

Liebl RA, Worsham AD (1987) Interference of Italian grass in wheat. Weed Sci 35:819–823

Lovett JV, Weerakoon WL (1983) Weed characteristics of the Labiatae, with special reference to allelopathy. Biol Agric Hortic 1:145–158

Maeder P, Hahn D, Dubois D, Gunst L, Alfoeldi T, Bergmann H, Oehme M, Amado R, Schneider H, Graf U, Velimirov A, Fliebbach A, Niggli U (2007) Wheat quality in organic and conventional farming: results of a 21 year field experiment. J Sci Food Agric 87:1826–1835

Mason H, Navabi A, Frick B, O'Donovan J, Niziol D, Spaner D (2007) Does growing Canadian Western Hard Red Spring wheat under organic management alter its breadmaking quality? Renewable Agric Food Syst 22:157–167

Mazzoncini M, Belloni P, Risaliti R, Antichi D (2007) Organic vs conventional winter wheat quality and organoleptic bread test. In: Proceedings of the 3rd international congress of the european integrated project "quality low input food" (QLIF). University of Hohenheim, Germany, pp 135–138

Nass H, Reiser B (1975) Grain filling period and grain yield relationships in spring wheat. Can J Plant Sci 55:673–678

Neacsu A, Serban G, Tuta C, Toncea I (2010) Baking quality of wheat cultivars, grown in organic, conventional and low input agricultural systems. Rom Agri Res 27:35–42

Niemeyer HM, Jerez JM (1997) Chromosomal location of genes for hydroxamic acid accumulation in *Triticum aestivum* L (wheat) using wheat aneuploids and wheat substitution lines. Heredity 79:10–14

Ohm JB, Chung OK (1999) Gluten, pasting, and mixograph parameters of hard winter wheat flours in relation to breadmaking. Cereal Chem 76:606–613

Olofsdotter M, Jensen LB, Courtois B (2002) Improving crop competitive ability using allelopathy: an example from rice. Plant Breeding 121:1–9

Preston KR, Hucl P, Townley-Smith TF, Dexter JE, Williams PC, Stevenson SG (2001) Effects of cultivar and environment on farinograph and Canadian short process mixing properties of Canada western red spring wheat. Can J Plant Sci 81:391–398

Pugsley A (1971) A genetic analysis of the spring-winter habit of growth in wheat. Crop Pasture Sci 22:21–31

Rao VV, Pandya SM (1992) Allelopathic influence of wheat on the growth of *Asphodelus tenuifalius*. In: Narwal SS, Tauro P (eds) Proceeding first national symposium on allelopathy in agroecosystems. Indian Society of Allelopathy, Haryana Agricultural University, Hisar, pp 41–44

Roberts D, Larson R (1985) Vernalization and photoperiodic responses of selected chromosome substitution lines derived from 'Rescue', 'Cadet', and 'Cypress' wheats. Can J Genet Cytol 27:586–591

Rodgers-Gray BS, Shaw MW (2000) Substantial reductions in winter wheat diseases caused by addition of straw but not manure to soil. Plant Pathol 49:590–599

Ryan MH, Derrick JW, Dann PR (2004) Grain mineral concentrations and yield of wheat grown under organic and conventional management. J Sci Food Agric 84:207–216

Santra DK, Santra M, Allan R, Campbell K, Kidwell K (2009) Genetic and molecular characterization of vernalization genes Vrn-A1, Vrn-B1, and Vrn-D1 in spring wheat germplasm from the Pacific Northwest Region of the USA. Plant Breeding 128:576–584

Sharma R (1994) Early generation selection for grain-filling period in wheat. Crop Sci 34:945–948

Shier NW, Kelman J, Dunson JW (1984) A comparison of crude protein, moisture, ash and crop yield between organic and conventionally grown wheat. Nutrition Reports Int 30:71–76

Shindo C, Tsujimoto H, Sasakuma T (2003) Segregation analysis of heading traits in hexaploid wheat utilizing recombinant inbred lines. Heredity 90:56–63

Sleper DA, Poehlman JM (2006) Breeding field crops. Blackwell publishing, Oxford

Sofield I, Evans L, Cook M, Wardlaw IF (1977) Factors influencing the rate and duration of grain filling in wheat. Funct Plant Biol 4:785–797

Stanciu G, Neacsu A (2008) Effects of genotype, nitrogen fertilizer and water stress on mixing parameters in wheat (*Triticum aestivum* L.). Rom Agri Res 25:29–35

Starling W, Richards MC (1993) Quality of commercial samples of organically grown wheat. Aspects Appl Biol 36:205–209

Stelmakh A (1993) Genetic effects of Vrn genes on heading date and agronomic traits in bread wheat. Euphytica 65:53–60

Stelmakh AF (1998) Genetic systems regulating flowering response in wheat (Reprinted from Wheat: Prospects for global improvement, 1998). Euphytica 100:359–369

Talbert L, Lanning S, Murphy R, Martin J (2001) Grain fill duration in twelve hard red spring wheat crosses. Crop Sci 41:1390–1395

Wang H, McCaig T, DePauw R, Clarke F, Clarke J (2002) Physiological characteristics of recent Canada western red spring wheat cultivars: yield components and dry matter production. Can J Plant Sci 82:299–306

Worland AJ (1996) The influence of flowering time genes on environmental adaptability in European wheats. Euphytica 89:49–57

Worland T, Snape J (2001) Genetic basis of worldwide wheat varietal improvement. The world wheat book: a history of wheat breeding. Lavoisier Publishing, Paris, pp 61–67

Worthington M, Reberg-Horton C (2013) Breeding cereal crops for enhanced weed suppression: optimizing allelopathy and competitive ability. J Chem Ecol 39:213–231

Wu H, Pratley J, Lemerle D, Haig T (2000) Evaluation of seedling allelopathy in 453 wheat (*Triticum aestivum*) accessions against annual ryegrass (*Lolium rigidum*) by the equal-compartment-agar method. Aust J Agric Res 51:937–944

Wu H, Pratley J, Ma W, Haig T (2003) Quantitative trait loci and molecular markers associated with wheat allelopathy. Theor Appl Genet 107:1477–1481

Wu HW, Haig T, Pratley J, Lemerle D, An M (1999) Simultaneous determination of phenolic acids and 2,4-dihydroxy-7-methoxy-1,4-benzoxazin-3-one in wheat (*Triticum aestivum* L.) by gas chromatography-tandem mass spectrometry. J Chromatogr A 864:315–321

Yan L, Helguera M, Kato K, Fukuyama S, Sherman J, Dubcovsky J (2004) Allelic variation at the VRN-1 promoter region in polyploid wheat. Theor Appl Genet 109:1677–1686

Zhang X, Xiao Y, Zhang Y, Xia X, Dubcovsky J, He Z (2008) Allelic variation at the vernalization genes, and in chinese wheat cultivars and their association with growth habit. Crop Sci 48:458–470

Chapter 5
Conclusion

Organic and low-input farming systems are extremely diverse with respect to environmental conditions, nutrients, moisture, diseases, insects/pests, and weed abundance. Cultivars developed for conventional management systems may not be suitable for organic systems. For example, different yield ranking of cultivars/lines in organic and conventional management systems suggests that indirect selection made during breeding on conventional lands may not result in the selection of best lines for organic management system. This implies that breeding wheat for organic agriculture should either be conducted on organic lands or care must be taken during selection process on conventional lands with respect to agronomic and allelopathic traits conferring competitive ability against weeds, diseases, nutrients, and water use efficiency. Considerable variation exist in the wheat germplasm to improve traits conferring competitiveness against weeds by breeding or agronomic techniques. Once a new wheat ideotype/cultivar with enhanced competitiveness is grown on a wider area, it is likely that weed species may be adapted to the changing conditions and become more competitive with wheat/crop plants as we have seen in case of herbicide resistant weeds. Herbicide resistance in many weed species has brought about a resurgence of interest among scientific (Table 5.1) and farming community to breed and cultivate wheat cultivars with improved competitive ability against weeds. However, it is recommended that wheat competitiveness should be considered as only one of the integrated weed management strategy. It is also suggested that reduction in herbicide use will not only help to manage the problem of herbicide resistant weeds but also help to improve crop competitiveness with optimum agronomic practices. The outcome of competitive ability between crop and weed plants is highly variable due to complex interaction of environment (climatic and edaphic), management (soil nutrient dynamics) and genetic (physiological, morphological, and biochemical traits) factors and thus, long-term studies are needed, especially in organic agriculture, to determine competitive ability and to study the effect of climate change on crop-weed relationship. In the last few decades, marker assisted breeding (MAB) has gained importance among plant breeders. Marker assisted breeding is an additional tool

M. Asif et al., *Managing and Breeding Wheat for Organic Systems*,
SpringerBriefs in Agriculture, DOI: 10.1007/978-3-319-05002-7_5,
© The Author(s) 2014

Table 5.1 Studies conducted during last five years to examine competitive ability in various cereal crops

S. No.	Crop	Studies/References
1	Wheat	(Ashinie et al. 2011; Behdarvand et al. 2012; Beres et al. 2010; Bernicot et al. 2010; Bertholdsson 2010; Carvalho et al. 2011; Drews et al. 2009; Fang et al. 2011; Feizabady et al. 2010; Fraga et al. 2013; Hamidi et al. 2010; Hassan et al. 2012; Hassannejad and Ghafarbi 2013; Hesammi 2011; Hooshyar et al. 2010; Khan et al. 2010; Lamego et al. 2013; Olsen et al. 2012; Reid et al. 2009a, b, 2011; Rigoli et al. 2009; Saadatian et al. 2010; Sandhu et al. 2010; Singh and Siya 2011, 2012; Singh et al. 2013; Szewczyk 2013; Travlos 2012; Worthington et al. 2013)
2	Rice	(Abdul Shukor et al. 2012; Aminpanah 2012a, b; Aminpanah and Javadi 2011; Anwar et al. 2010; Binang et al. 2011; Chauhan and Johnson 2010a, b; Dai et al. 2014; Dal Magro et al. 2011; Mahajan and Chauhan 2011; Md. Parvez et al. 2010; Meksawat and Pornprom 2010; Mennan et al. 2011, 2012; Moukoumbi et al. 2011; Namuco et al. 2009; Pheng et al. 2009; Rodenburg et al. 2009; Saito 2010; Saito et al. 2010; Shimizu and Itoh 2012; Singh Chauhan 2013; Tang et al. 2009; Tironi et al. 2009; Walia et al. 2010; Westendorff et al. 2013; Zhao et al. 2009)
3	Oat	(Lehnhoff et al. 2013; Schaedler et al. 2009)
4	Barley	(Auskalniene et al. 2010; Corre-Hellou et al. 2011; Dhima et al. 2010; Galon et al. 2011; Hamidi et al. 2010; Kokare and Legzdina 2010; Pilipavicius et al. 2011)
5	Maize	(Cury et al. 2012; Karimmojeni et al. 2010; Simic et al. 2009, 2012; So et al. 2009; Takim 2012; Wandscheer et al. 2013; Zadeh et al. 2011)

that offer great opportunity to make selection response more efficient. Various gene/QTLs have been identified (Table 3.1), mapped, and molecular markers have been developed for various agronomic and allelopathic traits coffering competitive ability against weeds which can be deployed in wheat breeding programs aimed to develop cultivars for organically managed and low-input farming systems.

References

Abdul Shukor J, Anwar MP, Ahmad S, Puteh A, Azmi M (2012) The influence of seed priming on weed suppression in aerobic rice. Pak J Weed Sci Res 18:257–264

Aminpanah H (2012a) Competition between rice (*Oryza sativa* L.) and barnyardgrass (*Echinochloa crus-galli* (L.) P. Beauv.) or rice barnyardgrass (*Echinochloa oryzicola* Vasing.). Rom Agr Res 29:395–403

Aminpanah H (2012b) Grain yield and nutrient uptake of rice (*Oryza sativa* L.) cultivars in competition with barnyardgrass *Echinochloa crus-galli* (L.) Beauv. Philippine Agr Sci 95:112–118

Aminpanah H, Javadi M (2011) Competitive ability of two rice cultivars (*Oryza sativa* L.) with barnyardgrass (*Echinochloa crus-galli* (L.) P.Beauv.) in a replacement series study. Adv Environ Biol 5:2669–2675

Anwar MP, Juraimi AS, Man A, Puteh A, Selamat A, Begum M (2010) Weed suppressive ability of rice (*Oryza sativa* L.) germplasm under aerobic soil conditions. Aust J Crop Sci 4:706–717

Ashinie B, Kedir N, Habtamu S (2011) Selection of some morphological traits of bread wheat that enhance the competitiveness against wild oat (*Avena fatua* L.). World J Agr Sci 7:128–135

Auskalniene O, Psibisauskiene G, Auskalnis A, Kadzys A (2010) Cultivar and plant density influence on weediness in spring barely crops. Zemdirbyste-Agriculture 97:53–60

Behdarvand P, Chinchanikar GS, Dhumal KN (2012) Influences of different nitrogen levels on competition between spring wheat (*Triticum aestivum* L.) and wild mustard (*Sinapis arvensis* L.). J Agr Sci (Toronto) 4:134–139

Beres BL, Harker KN, Clayton GW, Bremer E, Blackshaw RE, Graf RJ (2010) Weed-competitive ability of spring and winter cereals in the Northern Great Plains. Weed Technol 24:108–116

Bernicot MH, Rolland B, Fontaine L, Lecuyer J (2010) Wheat varieties in competition with weeds for sustainable agriculture, in particular organic farming. In: Goldringer I, Dawson JC, Rey F, Vettoretti A, Chable V, Lammerts van Bueren E, Finckh M, Barot S (eds) Breeding for resilience: a strategy for organic and low-input farming systems? EUCARPIA 2nd conference of the organic and low-Input agriculture section, Paris, 1–3 December 2010, pp 80–83

Bertholdsson NO (2010) Breeding spring wheat for improved allelopathic potential. Weed Res 50:49–57

Binang WB, Ekeleme F, Ntia JD (2011) Management of weeds of rainfed lowland rice using cultivar mixture strategies. Asian J Agri Res 5:306–311

Carvalho LBd, Alves PLdCA, Martins JVF (2011) Effects of plant density and proportion on the interaction between wheat with Alexandergrass plants. Bragantia 70:40–45

Chauhan BS, Johnson DE (2010a) Relative importance of shoot and root competition in dry-seeded rice growing with junglerice (*Echinochloa colona*) and ludwigia (*Ludwigia hyssopifolia*). Weed Sci 58:295–299

Chauhan BS, Johnson DE (2010b) Responses of rice flatsedge (*Cyperus iria*) and barnyardgrass (Echinochloa crus-galli) to rice interference. Weed Sci 58:204–208

Corre-Hellou G, Dibet A, Hauggaard-Nielsen H, Crozat Y, Gooding M, Ambus P, Dahlmann C, von Fragstein P, Pristeri A, Monti M, Jensen ES (2011) The competitive ability of pea-barley intercrops against weeds and the interactions with crop productivity and soil N availability. Field Crops Res 122:264–272

Cury JP, Santos JB, Silva EB, Byrro ECM, Braga RR, Carvalho FP, Valadao Silva D (2012) Accumulation and partitioning of dry matter and nutrients in maize cultivars in competition with weed. Planta Daninha 30:287–296

Dai L, Dai W, Song X, Lu B, Qiang S (2014) A comparative study of competitiveness between different genotypes of weedy rice (*Oryza sativa*) and cultivated rice. Pest Manag Sci 70:113–122

Dal Magro T, Schaedler CE, Fontana LC, Agostinetto D, Vargas L (2011) Competitive ability between biotypes of *Cyperus difformis* L. resistant or susceptible to ALS-inhibitor herbicide and those with flooded rice. Bragantia 70:294–301

Dhima K, Vasilakoglou I, Gatsis T, Eleftherohorinos I (2010) Competitive interactions of fifty barley cultivars with *Avena sterilis* and *Asperugo procumbens*. Field Crops Res 117:90–100

Drews S, Neuhoff D, Koepke U (2009) Weed suppression ability of three winter wheat varieties at different row spacing under organic farming conditions. Weed Res 49:526–533

Fang Y, Liu L, Xu B-C, Li F-M (2011) The relationship between competitive ability and yield stability in an old and a modern winter wheat cultivar. Plant Soil 347:7–23

Feizabady AZ, Sareban H, Khazaei H (2010) Effect of different densities of wild oat (*Avena ludovicianal*) on wheat cultivars yield. In: Proceedings of 3rd Iranian weed science congress, vol 1, Weed biology and ecophysiology, Babolsar, Iran, 17–18 February 2010, pp 269–273

Fraga DS, Agostinetto D, Vargas L, Nohatto MA, Thuermer L, Holz MT (2013) Adaptive value of ryegrass biotypes with low-level resistance and susceptible to the herbicide fluazifop and competitive ability with the wheat culture. Planta Daninha 31:875–885

Galon L, Tironi SP, Rocha PRR, Concenco G, Silva AF, Vargas L, Silva AA, Ferreira EA, Minella E, Soares ER, Ferreira FA (2011) Competitive ability of barley cultivars against ryegrass. Planta Daninha 29:771–781

Hamidi R, Mazaheri D, Rahimian H, Alizadeh HM, Zeinali H (2010) Effects of nitrogen and population density on wild barley (*Hordeum spontaneum* Koch) competition with winter wheat (*Triticum aestivum* L.). In: Proceedings of 3rd Iranian weed science congress, vol 1, Weed biology and ecophysiology, Babolsar, Iran, 17–18 February 2010, pp 335–338

Hassan IAK, Khan SA, Shah SMA (2012) Wheat-wild oats interactions at varying densities and proportions. Pak J Bot 44:1053–1057

Hassannejad S, Ghafarbi SP (2013) Distribution of dominant weed species in winter wheat at Tabriz county. Int J Biosci 3:8–16

Hesammi E (2011) Evaluating competition of the Phalaric minor in wheat. J Am Sci 7:342–346

Hooshyar R, Zand E, Bagestani MA, Razzazi A (2010) Study of competitive ability of seven genotypes of temperate zone wheat with rye. In: Proceedings of 3rd Iranian weed science congress, vol 1, Weed biology and ecophysiology, Babolsar, Iran, 17–18 February 2010, pp 339–341

Karimmojeni H, Mashhadi HR, Shahbazi S, Taab A, Alizadeh HM (2010) Competitive interaction between maize, *Xanthium strumarium* and *Datura stramonium* affecting some canopy characteristics. Aust J Crop Sci 4:684–691

Khan IA, Hassan G, Marwat KB, Daur I, Shah SMA, Khan NU, Khan SA, Farhatullah (2010) Interaction of wild oat (*Avena fatua* L.) with divergent wheat cultivars. Pak J Bot 43:1051–1056

Kokare A, Legzdina L (2010) Traits influencing spring barley competitiveness against weeds under organic and conventional conditions. In: Annual 16th international scientific conference proceedings, "Research for rural development 2010", Jelgava, Latvia, vol 1, pp 11–16, 19–21 May 2010

Lamego FP, Ruchel Q, Kaspary TE, Gallon M, Basso CJ, Santi AL (2013) Competitive ability of wheat cultivars against weeds. Planta Daninha 31:521–531

Lehnhoff EA, Keith BK, Dyer WE, Menalled FD (2013) Impact of biotic and abiotic stresses on the competitive ability of multiple herbicide resistant wild oat (*Avena fatua*). Plos One 8:e64478

Mahajan G, Chauhan BS (2011) Effects of planting pattern and cultivar on weed and crop growth in aerobic rice system. Weed Technol 25:521–525

Md. Parvez A, Abdul Shukor J, Azmi M, Adam P, Ahmad S, Mahfuza B (2010) Weed suppressive ability of rice (*Oryza sativa* L.) germplasm under aerobic soil conditions. Aust J Crop Sci 4:706–717

Meksawat S, Pornprom T (2010) Allelopathic effect of itchgrass (*Rottboellia cochinchinensis*) on seed germination and plant growth. Weed Biol Manag 10:16–24

Mennan H, Kaya-Altop E, Ngouajio M, Sahin M, Isik D (2011) Allelopathic potentials of rice (*Oryza sativa* L.) cultivars leaves, straw and hull extracts on seed germination of barnyardgrass (*Echinochloa crus-galli* L.). Allelopathy J 28:167–178

Mennan H, Ngouajio M, Sahin M, Isik D, Altop EK (2012) Competitiveness of rice (*Oryza sativa* L.) cultivars against *Echinochloa crus-galli* (L.) Beauv. in water-seeded production systems. Crop Prot 41:1–9

Moukoumbi YD, Sie M, Vodouhe R, Bonou W, Toulou B, Ahanchede A (2011) Screening of rice varieties for their weed competitiveness. Afr J Agric Res 6:5446–5456

Namuco OS, Cairns JE, Johnson DE (2009) Investigating early vigour in upland rice (*Oryza sativa* L.): Part I. Seedling growth and grain yield in competition with weeds. Field Crops Res 113:197–206

Olsen JM, Griepentrog H-W, Nielsen J, Weiner J (2012) How important are crop spatial pattern and density for weed suppression by spring wheat? Weed Sci 60:501–509

Pheng S, Olofsdotter M, Jahn G, Adkins SW (2009) Potential allelopathic rice lines for weed management in Cambodian rice production. Weed Biol Manag 9:259–266

Pilipavicius V, Romaneckiene R, Romaneckas K (2011) Crop stand density enhances competitive ability of spring barley (*Hordeum vulgare* L.). Acta Agri Scand 61:648–660

Reid TA, Navabi A, Cahill JC, Salmon D, Spaner D (2009a) A genetic analysis of weed competitive ability in spring wheat. Can J Plant Sci 89:591–599

Reid TA, Yang R-C, Salmon DF, Navabi A, Spaner D (2011) Realized gains from selection for spring wheat grain yield are different in conventional and organically managed systems. Euphytica 177:253–266

Reid TA, Yang R-C, Salmon DF, Spaner D (2009b) Should spring wheat breeding for organically managed systems be conducted on organically managed land? Euphytica 169:239–252

Rigoli RP, Agostinetto D, Vaz Da Silva JMB, Fontana LC, Vargas L (2009) Competitive potential of wheat cultivars as affected by emergence time. Planta Daninha 27:41–47

Rodenburg J, Saito K, Kakai RG, Toure A, Mariko M, Kiepe P (2009) Weed competitiveness of the lowland rice varieties of NERICA in the southern Guinea Savanna. Field Crops Res 114:411–418

Saadatian B, Soleymani F, Ahmadvand G (2010) Effect of rye (Secale cereale) competition on growth indices and yield of two wheat (*Triticum aestivum*) cultivars. In: Proceedings of 3rd Iranian weed science congress, vol 1, Weed biology and ecophysiology, Babolsar, Iran, 17–18 February 2010, pp 297–300

Saito K (2010) Weed pressure level and the correlation between weed competitiveness and rice yield without weed competition: an analysis of empirical data. Field Crops Res 117:1–8

Saito K, Azoma K, Rodenburg J (2010) Plant characteristics associated with weed competitiveness of rice under upland and lowland conditions in West Africa. Field Crops Res 116:308–317

Sandhu RS, Gill BS, Chahal PS (2010) Competitive ability of wheat (*Triticum aestivum* L.) against wild oats (*Avena ludoviciana* Dur.) as influenced by date of sowing, seed rate and spacing. Indian J Ecol 37:97–100

Schaedler CE, Fleck NG, Ferreira FB, Lazaroto CA, Rizzardi MA (2009) Morphological traits in oat plants cultivars as indicators of competitive potential against weeds. Ciencia Rural 39:1313–1319

Shimizu H, Itoh Y (2012) Developmental change of vegetative plant architecture of annual-form-wild rice (*Oryza rufipogon* Griff.) elevates competitive ability during the late development under a dense condition. Am J Plant Sci 3:670–687

Simic M, Dolijanovic Z, Maletic R, Filipovic M, Grcic N (2009) The genotype role in maize competitive ability. Genetika-Belgrade 41:59–67

Simic M, Dolijanovic Z, Maletic R, Stefanovic L, Filipovic M (2012) Weed suppression and crop productivity by different arrangement patterns of maize. Plant Soil Environ 58:148–153

Singh Chauhan B (2013) Phenotypic plasticity of blistering ammannia (*Ammannia baccifera*) in competition with direct-seeded rice. Weed Technol 27:373–377

Singh RK, Siya R (2011) Competitive ability of wheat cultivars against complex weed flora in zero-till planting. In: 23rd Asian-Pacific weed science society conference vol 1, weed management in a changing world, Cairns, Queensland, Australia, 26–29 September 2011, pp 537–543

Singh RK, Siya R (2012) Competitive ability of wheat cultivars against complex weed flora in zero-till planting. Pak J Weed Sci Res 18:629–635

Singh V, Singh H, Raghubanshi AS (2013) Competitive interactions of wheat with *Phalaris minor* or *Rumex dentatus*: a replacement series study. Int J Pest Manag 59:245–258

So YF, Williams MM II, Pataky JK, Davis AS (2009) Principal canopy factors of sweet corn and relationships to competitive ability with wild-proso millet (*Panicum miliaceum*). Weed Sci 57:296–303

Szewczyk BF (2013) The influence of morphological features of spelt wheat (*Triticum aestivum* ssp spelta) and common wheat (*Triticum aestivum* ssp vulgare) varieties on the competitiveness against weeds in organic farming system. J Food Agric Environ 11:416–421

Takim FO (2012) Weed competition in maize (Zea mays L.) as a function of the timing of hand-hoeing weed control in the southern guinea savanna zone of Nigeria. Acta Agron Hung 60:257–264

Tang J, Xie J, Chen X, Yu L (2009) Can rice genetic diversity reduce *Echinochloa crus-galli* infestation? Weed Res 49:47–54

Tironi SP, Galon L, Concenco G, Ferreira EA, Silva AF, Aspiazu I, Ferreira FA, Silva AA, Noldin JA (2009) Competitive ability of rice plants with barnyardgrass biotypes resistant or susceptible to quinclorac. Planta Daninha 27:257–263

Travlos IS (2012) Reduced herbicide rates for an effective weed control in competitive wheat cultivars. Int J Plant Prod 6:1–13

Walia US, Gill G, Walia SS, Sidhu AS (2010) Competitive ability of rice genotypes against weeds in direct seeding production system. Ind J Weed Sci 42:30–34

Wandscheer ACD, Rizzardi MA, Reichert M (2013) Competitive ability of corn in coexistence with goosegrass. Planta Daninha 31:281–289

Westendorff NR, Agostinetto D, Ulguim AR, Langaro AC, Thuermer L (2013) Initial growth and competitive ability of yellow nutsedge and irrigated rice. Planta Daninha 31:813–821

Worthington ML, Reberg-Horton SC, Jordan D, Murphy JP (2013) A comparison of methods for evaluating the suppressive ability of winter wheat cultivars against italian ryegrass *(Lolium perenne)*. Weed Sci 61:491–499

Zadeh HG, Lorzadeh S, Aryannia N (2011) Evaluating weeds competitive ability in a corn field in southern west of Iran. Asian J Crop Sci 3:179–187

Zhao DL, Atlin GN, Bastiaans L, Spiertz JHJ (2009) Opportunities for ecologically-based aerobic rice cropping systems: weed competitiveness of genotypes. In: Kingely RV (ed) Weeds: Management, economic impacts and biology, pp 23–39

Author Biographies

Muhammad Asif was employed with Pakistan Agricultural Research Council, Islamabad, Pakistan as Research Scientist since 1996 where he was involved in the development of high-yielding, drought-tolerant, and disease-resistant wheat cultivars through conventional breeding. He completed his B.Sc. (Hons.) and M.Sc. (Hons.) with specialization in Plant Breeding and Genetics from University of Arid Agriculture, Rawalpindi, Pakistan. He has written more than 20 scientific research/review articles, 10 book chapters, and edited/authored four books. He immigrated to Canada during 2009. Over the last 4 years at Department of Agricultural, Food and Nutritional Science, University of Alberta, he has focused on the improvement of doubled haploid production technology in wheat and triticale along with QTL mapping and marker-assisted breeding for various agronomic, quality, disease resistance, and competitive ability traits in wheat.

Muhammad Iqbal obtained his B.Sc. (Hons) and M.Sc. (Hons) degrees in Agriculture with specialization in Plant Breeding and Genetics from Agricultural University, Peshawar, Pakistan. He then joined Pakistan Agricultural Research Council, Islamabad in 2001 as Research Scientist. He studied at the University of Alberta, Edmonton, Canada and completed his PhD studies in December 2006. Afterward, he worked in Plant Biotechnology Program, National Agricultural Research Center (NARC), Islamabad, Pakistan for 6 years. He also taught at the Department of Plant Genomics and Biotechnology, Pakistan Agricultural Research Council's Institute of Advanced Studies in Agriculture, NARC, Islamabad for 6 years. He has supervised/cosupervised the thesis research of 15 M.Phil. and 5 Ph.D. students. Presently, he is a postdoctoral Fellow at the University of Alberta, Edmonton, Canada. Dr. Iqbal has worked on the genetic improvement of maize, Rice, Papaya, and Wheat. His research interests include quantitative trait loci mapping and marker-assisted selection for various agronomic and quality traits and disease resistance in crops, especially wheat. He has published more than 30 research/review articles in journals of international repute.

M. Asif et al., *Managing and Breeding Wheat for Organic Systems*,
SpringerBriefs in Agriculture, DOI: 10.1007/978-3-319-05002-7,
© The Author(s) 2014

Harpinder Singh Randhawa is a spring wheat breeder working at Agriculture and Agri-Food Canada. His passion for wheat breeding developed during his childhood while he grew up and worked alongside his parents on the family farm in Punjab. He obtained his B.Sc. Agriculture (Hons) in 1990 and M.Sc. with a specialization in Plant Breeding in 1993 from Punjab Agricultural University, Ludhiana. The focus of his research was to improve the bread-making quality of wheat using novel genes from wheat progenitors. In 1994, he was appointed as Assistant Rice Breeder at the Punjab Agricultural University where he was part of a team whose objective was to develop high-yielding cultivars of rice. He joined his PhD program at the University of Saskatchewan, Saskatoon, in 1998 where he conducted genetic and molecular studies of loose smut resistance in durum wheat. In 2002, he became a postdoctoral Fellow at the University of Nebraska, Lincoln and later at Washington State University, Pullman. Dr. Randhawa focused his research to develop new wheat genomics tools, novel strategies for rapid introgression of traits using marker-assisted backcrossing, genetic, and physical mapping of agronomically important traits in wheat and eventually developing improved wheat cultivars. Since 2007, Dr. Randhawa has been working as a spring wheat and triticale breeder with Agriculture and Agri-Food Canada at the Lethbridge Research Centre, Alberta. His prime focus of research is to develop spring wheat cultivars that are high yielding with excellent end use quality and resistance to various biotic and abiotic stresses in western Canada. He has developed five high-yielding spring wheat cultivars and codeveloped three high-yielding triticale cultivars for general production in western Canada. His other research interests include the identification of new sources of disease resistance in wheat, genetic mapping, doubled haploid production, and new breeding tools. He has published over 36 research articles in the international journals, supervised many undergraduate and graduate students and postdoctoral fellows, and attended over 25 National and International conferences.

Dean Spaner is a Professor and spring wheat breeder in the Department of Agricultural, Food and Nutritional Science, University of Alberta, Edmonton, Alberta, Canada. His research group works in the areas of wheat breeding and genetics, and on wheat breeding, genetics and agronomy for organic environments.

Index

A

Aegilops speltoides, 3
Aegilops squarrosa, 3
Aegilops tauschii, 3, 34
Albumins, 4
Allelopathy, 55–57
Allohexaploid, 3
Alopecurus myosuroides, 15
Amplified fragment length polymorphisms
    (AFLPs), 44
Annual rye grass (*Lolium rigidum* L.), 11, 23,
    32, 55, 56
Arbuscular mycorrhizas (AMs), 34
Area, 2f, 9
    canopy area, 16
    under organic agriculture, 10
Avena spp., 11

B

Baking/bread quality, 54
Barley (*Hordeum vulgare*), 10, 14, 15, 22, 23,
    24, 28, 66t
    competitive ability, 66t
    competitiveness, 31
    crop mixture, 30
    crop rotation, 28
    genotypes, 55, 56
Benzoxazinoids (BX), 56
Biological control, 22
Biomass
    crop, 14, 55
    plant, 16
    seedling, 15
    weed, 15, 16, 17, 23, 24, 25, 26, 30, 32, 56
Bran, 4
Brassica hirta, 24
Bread wheat. *See* Wheat (*Triticum aestivum*)
Breeding
    for allelopathy, 55–57
    for early maturity, 57–61
    for quality, 54–55
    resistant cultivars, 54
Buckwheat (*Fagopyrum esculentum*), 22
Bunts, 3, 53

C

Canada Prairie Spring Wheat (CPSW) class,
    31
Canada Seeds Act (CSA), 22
Canada Western Red Spring (CWRS) class,
    31, 32
Canola (*Brassica* sp.), 10, 22
Canopy, 12, 14, 15, 24
    crop canopy development, 16, 26
    overall development, 17
Chenopodium album L., 35
Chromosome pairing, 3
Chrysolina hyperici (leaf beetle), 22
Chuan–Mai18 (CM18), 33
Cirsium arvense (Canada thistle), 22
Colletotrichum gloeosporioides, 22
Competitive ability, 12, 13, 13f, 14, 15, 16.
        See also Crop competitiveness
Conventional management system, 12, 31, 45,
    53, 54, 55, 65
    early stage wheat in, 25f
    late stage wheat in, 26f
Cover crops, 26–28
Crop competitiveness
    association of plant traits to, 13–17
    biotic and abiotic factors, 13f
    definition, 12
    genetics and genomics of, 31–35, 44–45
    importance of, 10–12
    management–related factors, 12
    order of, 13f

M. Asif et al., *Managing and Breeding Wheat for Organic Systems*,   73
SpringerBriefs in Agriculture, DOI: 10.1007/978-3-319-05002-7,

Crop competitiveness (*cont.*)
  plant–related factors, 12
  weed–related factors, 12
Crop density, 22, 23–24, 53
Crop morphology, 14, 16, 17
Crop rotation, 28–30
Cultural control, 22
  cover crops (*see* Cover crops)
  crop density (*see* Crop density)
  crop or cultivar mixtures, 30
  crop rotation (*see* Crop rotation)
  fertilization, 24
  mechanical weed control, 30–31
  row spacing and planting pattern, 24–26

D
Density, 12, 14, 24, 28
  crop density (*see* Crop density)
Deoxyribonucleic acid (DNA), 3, 35, 44, 58
Disease resistance, 53–54
Doubled haploid (DH) line, 33
Downy brome (*Bromus tectorum*), 31
Drechslera tritici repentis, 54

E
Early maturity, 57–61
  delayed planting, 57
  in spring wheat, 57
Early season vigor, 14–15
Endosperm proteins, 4

F
Federal Plant Protection Act (FPPA), 22
Flax (*Linum usitatissimum*), 22
Food and Agriculture Organization (FAO), 10
Fusarium Head Blight (FHB), 3, 53
Fusarium graminearum, 3

G
Galium aparine, 35
Genetically modified organisms (GMO), 10
Germ, 4
Gliadins, 4
Gluten, 4
Glutenins, 4
Goat grass (*Aegilops cylindrica*), 16
Grain fill duration (GFD), 60
Grain fill rate (GFR), 60
Grain protein content, 4

Grain texture, 4
Grain yield, 60
Granule–bound starch synthase. *See* Waxy
  (Wx) proteins
Green Revolution, 3, 10, 11, 15

H
Ha locus, 4
Haploid nuclear genome, 3
Hardness, 1, 4
Herbicides, 10, 11, 12, 35, 65
  synthetic, 21
Hypericum perforatum L., 22

I
Identifying/pyramiding genes, 44
Integrated weed management strategies
  (IWMS), 12
Interference, 12
Interspecific competition, 12
Intraspecific competition, 12, 23

L
Lamiuma plexicaule (henbit), 23
Leaf area index (LAI), 14, 15, 16
Leaf rust (*Puccinia triticina*), 3
Lentils (*Lens culinaris*), 22
Light interception, 16–17
Linum multiflorum, 35
Loose smut (*Ustilago tritici*), 3
Low–input farming systems, 65

M
Maize, 1, 28
  competitive ability, 66*t*
Marker assisted breeding (MAB), 35, 65
Microsatellites, 44
Milling, 4
Modifiers of earliness, 59
Molecular markers, 35, 44, 45, 65
Multilocation testing, 33
Mustard (*Sinapis alba*), 55

N
Nitrogen use efficiency (NUE), 34, 44
Noxious Weed Control Act
  (NWCA), 22
Nutrients (N, P, K), 24

O

Oat (*Avena sativa*), 22
  competitive ability, 66*t*
Organic agriculture, 9, 10
Organic farming systems, 65

P

Peas (*Pisum sativum*), 10
Perennial rye grass (*Lolium perenne*), 55, 56
Ph1 (Pairing homoeologous) gene, 3
Phalaris brachystachys, 35
Phalaris minor, 35
Phalaris paradoxa, 23
Phoma macrostoma (fungus), 22
Photoperiod (Ppd), 57, 59
Photosynthetic active radiation (PAR), 14
Plant growth promoting rhizosphere (PGPR), 34
Plant height, 15, 16
Polymerase chain reaction (PCR), 44
  based DNA markers, 58
Powdery mildew (*Blumeria graminis*), 3, 53
Production, 1, 2*f*, 3
Prolamines, 4
Protein, 1, 2, 26
  concentration, 60
  endosperm protein, 4
  gluten, 4
  grain protein, 4, 54, 55, 60
  waxy protein, 5
Puroindoline a (*Pina*) gene, 4
Puroindoline b (*Pinb*) gene, 4

Q

Quality, 4, 11, 35, 44, 45, 60
  breeding for, 54–55
Quantitative trait loci (QTL), 33
  conferring competitive ability, 36–43*t*

R

Radiation use efficiency (RUE), 17
Random amplified polymorphic DNAs (RAPDs), 44
Raphanus raphanistrum, 11
Recombinant inbred line (RIL) population, 33
Relative emergence time, 14

Restriction fragment length polymorphisms (RFLPs), 44
Retrotransposons, 44
Rhizosphere competence, 34
Rht (height reducing) genes, 15
RhtB1b dwarfing gene, 33
Rice (*Oryza sativa*), 1, 10, 16, 41*t*, 42*t*, 43*t*, 66*t*
  biomass, 17
  competitive ability, 66*t*
  crop rotation, 28
Root rot (*Cochliobolus sativus*), 3
Rusts, 53
Rye (*Lolium* sp.), 22

S

Septoria nodorum, 54
Septoria tritici blotch, 54
Simple sequence repeats (SSRs), 44
Sinapis arvensis L., 35
Single nucleotide polymorphisms (SNPs), 44
Smuts, 53
Starch, 2, 5
  amylopectin, 5
  amylose, 5
  free starch, 4
Stripe rust (*Puccinia striiformis* f. sp. *tritici*), 3

T

Tillering, 11, 14, 16, 17, 23, 33
Triticum monococcum, 3
Triticum turgidum, 2
Triticum turgidum ssp. *dicoccum*, 3
Troublesome grassy weeds, 11

U

United Nations (UN), 10
United States (US), 22

V

Vernalization (Vrn), 57, 58, 59

W

Waxy (Wx) proteins, 5
  genes, 5

Weed control management strategies, 21–22
   preventive measures, 22
   conventional management, 25*f*, 26*f*
   organic management system, 25*f*, 27*f*
Weeds, 11, 11*f*, 12
Wheat (*Triticum aestivum*), 1, 10, 22, 23
   classes, 1
   growth habits, 2
   wheat producing countries, 2*f*
   history, 3
   production, 3
   utilization, 4
   crop mixture, 30
   crop rotation, 28
   competitive ability, 66*t*
Wheat stem rust (*Puccinia graminis* f. sp.
   tritici), 3

White mustard (*Sinapis alba*), 35
Wild oat (*Avena sterilis*), 11, 23, 35
Wild onion (*Asphodelus tenuifolius*), 56
Wild wheat (*Aegilops tauschii*), 3
World Health Organization (WHO), 10

X
Xanthomonas campestris pv. *poae*
   (bacterium), 22

Y
Yield, 1, 2*f*, 3, 12, 23, 26
   and plant height, 15